INTRODUCTION TO CULTURAL MATHEMATICS

INTRODUCTION TO CULTURAL MATHEMATICS

With Case Studies in the Otomies and Incas

THOMAS E. GILSDORF

A JOHN WILEY & SONS, INC., PUBLICATION

Published by John Wiley & Sons, Inc., Hoboken, New Jersey
Published simultaneously in Canada

Library of Congress Cataloging-in-Publication Data:

Gilsdorf, Thomas E.
Introduction to cultural mathematics: with case studies in the Otomies and Incas / Thomas E. Gilsdorf.
p. cm.
Includes bibliographical references and index.
ISBN 978-1-118-11552-7 (hardback)
1. Ethnomathematics. 2. Otomi Indians–Mathematics–Case studies.
3. Incas–Mathematics–Case studies. 4. Otomi Indians–Social life and customs.
5. Incas–Social life and customs. I. Title.
GN476.15.G55 2012
510–dc23

2011036452

Printed in the United States of America

ISBN: 978-1-118-11552-7

10 9 8 7 6 5 4 3 2 1

To Elisa, Inez, and Sophia

CONTENTS

PREFACE

Dear Reader,

You are about to see mathematics in a way that is perhaps quite different from how you have previously seen it. Cultural mathematics has to do with how people in different places and situations think about and create mathematics. Mathematics has much in common with language. There are interpretations and expressions of mathematics that vary across cultures, just as there are interpretations and expressions of language that vary across cultures. This book is about understanding and describing some of the variations of interpretations and expressions of mathematics. Here is what this means to you, the reader:

- This book is about mathematics as you perceive it, and as others perceive it. For example, you will be asked to think about and/or discuss what the mathematics of art, kinship, calendars, games, and other topics mean to you.
- This book is about discovering how different people can and do interpret mathematics in ways that are quite different from what you might be used to. I think you will find that humans are very creative about how to create and understand mathematics.
- This book is not about memorizing formulas and equations. There are plenty of mathematics courses and textbooks that emphasize that aspect of mathematics.
- This book is not about finding "the right answer." It is about understanding that there can be more than one way to perceive mathematics. In

many situations, there are many possible interpretations, so this is a chance for you to think creatively about mathematics!

I hope you enjoy this book and that you learn something new about connections between mathematics and culture.

Sincerely,
THOMAS E. GILSDORF

Grand Forks, North Dakota
October 2011

INTRODUCTION

Regardless of how you view mathematics, whether you like it or not, whether you feel you could explain what "mathematics" is or not, most people would agree that mathematics is a way of communicating information. That information may seem quite different from information that is communicated orally or in writing, yet it is still information of some kind. As a form of communication, mathematics is subject to influences and variations in interpretation both at the societal/cultural level and even at the individual level. In other words, our understanding of what is or is not mathematics is influenced both by what is perceived to be mathematics by the culture we live in and by what is perceived to be mathematics by each person in that culture. The interaction between mathematics and culture is what this book is all about. My choice of the expression "cultural mathematics" in the title comes from my intention that prospective readers know right from the start that the two main subjects are mathematics and culture. This topic also goes by other names that you may have heard about or read about. Cultural mathematics is also called *ethnomathematics*, a term first defined by Ubiratan D'Ambrosio in the 1980s. D'Ambrosio is considered to be the person who made the study of the interaction between mathematics and culture into an established discipline of study. Other expressions associated with the study of mathematics and culture include *multicultural mathematics*, *traditional mathematics*, *indigenous mathematics*, and *oral and written mathematics*.

My goal in writing this book is to make available a resource that could serve as a textbook for courses in which there is a significant emphasis on the connections between culture and mathematics. Although there are many excellent sources of information on the cultural aspects of mathematics, my idea is to include some course-like structure with my exposition, such as exercises and topics for essay/discussion. I have attempted to write this for use as the main text for a course on cultural mathematics (or ethnomathematics), or as a main or supplementary text for courses on mathematics education or multicultural education, for graduate courses on the connections between education and culture, for use as a supplementary text for courses in the history of mathematics, or as a general reference for anyone who is interested in the connections between mathematics and culture. This book is roughly at the level of someone who has had college algebra; however, the most important prerequisite for a reader is an interest in discovering connections between mathematics and culture.

I believe the beauty of mathematics should be enjoyed because of the human creativity that went into making it. After all, mathematics is very much a human endeavor. Unfortunately, much of modern mathematics is presented in a way that disconnects it from the fact that it has been created by people. What makes cultural mathematics so interesting to me is that the human aspect of it is *unavoidable* and, hence, plays a central role. The mathematics does not have to be confusing or esoteric to be beautiful. One of my main goals has been to maintain this attitude throughout the book. For those who seek cultural mathematics that does require deeper (Western) mathematical experience, I have included such mathematics in some of the exercises marked with an asterisk (*), and in the notes at the end of each chapter I have indicated where a person can investigate topics further.

Writing about the topic of cultural mathematics for readers with backgrounds primarily in Western mathematics brings one to a dilemma: On one hand, using Western terminology and notation to describe mathematics of non-Western cultures is inherently inaccurate because people in such cultures would not think of the mathematical content in the same way as it is perceived in Western culture. On the other hand, if the goal is for people of Western backgrounds to understand how cultural activities can be understood as mathematics, then one must speak to readers in familiar mathematical terms.

Thus, it is crucial that the reader has a foundation that includes an understanding of what "culture" means, what constitutes "mathematics," how a person's culture affects and often blurs that person's perceptions of mathematics, and how to keep aware of cultural tendencies and perceptions as one tries to understand the mathematics of other cultures. The goal of Chapter 1 is to put together such a foundation. In my effort to create it, I have rewritten Chapter 1 several times. I have numerous "half baked" word processing files to prove it! The final, "fully baked" version of Chapter 1 is much longer than I first imagined it to be, but that chapter is, without question, the most important chapter of the book.

I would like to briefly discuss some peripheral considerations that are beyond the intentions of this book. One such topic consists of the political and social aspects of cultural mathematics. These two aspects have to do with how a culture's political or social situation affects perceptions (by all cultures involved) of mathematics, how political and/or social events have affected (often adversely) the preservation of cultural knowledge, including mathematical knowledge, and the role mathematics (or science) plays in social/political contexts. Examples could include the effects on mathematics of one culture ruling over another culture, or DNA testing to determine cultural backgrounds (plus the mathematics related to such contexts). I believe these political and social aspects of mathematics are important considerations that relate to culture; however, I am not an expert in either of these areas, and so I have not tried to address them here. On the other hand, I hope this book will be useful to those who do have more expertise in these areas, and it would be great if some of those readers would be motivated to create scholarly works in the areas of social and political aspects of cultural mathematics.

The structure of the book is as follows: Part I, which consists of Chapters 1–8, is a description of the general ideas of cultural mathematics; and Part II, consisting of Chapters 9 and 10, represents two "case studies," in which I apply the ideas of Part I to a specific culture. The two cultures I chose, the Otomies of Mexico and the Incas of South America, are the two cultural groups I know best in terms of cultural mathematics. Exercises are included at the end of each chapter, followed by further notes about the topics of that chapter.

I have divided the exercises into three categories. The first type of exercise is *short answer*. In these questions I try to prod the reader to contemplate ideas related to the topic at hand. Second are *calculations*, in which the reader can practice the relevant concepts. In some cases the process is similar to how people in the cultural group at hand would have solved the problem. In other cases the process has been restated in Western mathematical terms. Working through the calculations this way gives the reader a feel for how the calculation is accomplished. The third type of exercise is *essay/discussion*, in which the reader can delve more deeply into a topic, or it could serve as a topic for group discussions. Whenever an exercise requires some specialized mathematical knowledge (e.g., at or beyond the level of calculus), I have marked it with an asterisk (*) to alert the reader.

At the end of the book are hints and comments about some of the exercises. I chose not to simply include numerical answers to some problems for essentially two reasons. One reason is that I do not believe in the statement, "If I get the correct answer, then I understand the concept." A second reason is that a course in cultural mathematics, being very different from a typical mathematical course, requires a broad state of thinking. There are many situations in which more than one solution or explanation is possible. There are others in which we do not know what the solution or explanation is. In some cases, I simply ask the reader to describe one of perhaps many possible explanations to the situation or concept being considered. On the other hand, I want readers to feel that they understand what the topic is about. Hopefully, the suggestions and comments will lead readers toward a better understanding of the topic relevant to the exercise.

In the bibliography, I have attempted to include as many references as I could think of so that interested readers can seek more information and/or study further.

ON USING THIS BOOK IN A CLASSROOM

If this book is used as a course textbook, I have some comments about its organization and such. Perhaps this is the first time you are teaching a course in cultural mathematics. In any case, I think you will find that it is quite different from teaching a traditional mathematics course.

Motivational questions: One of the most exciting aspects of learning about cultural mathematics is that a person gets to think about mathematics in a very creative way. In my own teaching experience, I have found that students start thinking creatively about mathematics if I ask them to describe their own interpretation of concepts before going through the material. The motivational questions at the beginning of each chapter are intended to mimic this. My suggestion to the instructor is to have the students explain verbally or write down on paper their responses to the motivational questions before you begin to go through the topics. If you feel that some students will not put much effort into responding to the questions, you could have their responses to the questions be worth a nominal number of points. Also, you may wish to mention to them that their responses should be based on how they perceive the topic, not on how they think it is "supposed to be." You will notice that during the rest of a particular chapter I discuss the issues that come up regarding the motivational questions; however, the intention of the questions is to have students think carefully about certain aspects of mathematics as they apply to their personal lives and experiences, and then, during the rest of the chapter, they can see how people who have other experiences view the same mathematical aspects.

The classroom setting: Another exciting aspect of learning about cultural mathematics is that it is an excellent opportunity to discover mathematical concepts. In order to set up that kind of environment, I have written the book in an informal style, with the idea that the classroom setting will be one in which everyone feels encouraged to think about mathematics and discover mathematical ideas, including you, the instructor. Below are some aspects of teaching a cultural mathematics course that I believe will enhance the experience for you and your students. My intention has been for this book to make the aspects easier to implement.

- Ask students to draw on their own experiences from their personal lives (such as family life and background, job or school experiences, or previous mathematical experiences).
- Keep in mind that the course is not about finding "correct answers"; there are many ways to interpret how humans do mathematics. Human

diversity manifests itself in mathematics just as it does in many other disciplines.

- As often as possible, give students an opportunity to discover the mathematics on their own. For example, most of the exercises can be given to students in class, the idea being that they work in groups or individually to come up with a response. Those responses could either be turned in and counted for points or discussed in class. Some exercises could be done in class and others done as assignments, for example.
- You do not need to be an expert on cultural mathematics in order to teach a course on the topics covered in this book. In fact, you should expect that from time to time students will ask about or comment on something for which you do not have full information (this happens to me frequently whenever I teach cultural mathematics). Such moments represent opportunities for students to learn and discover on their own. You could ask the students—or, better yet, the whole class—to look up references in the bibliography of this book, to go to the library, to arrange to interview someone who knows about the topic, and so forth, in order to understand the topic.

Suggested chapters to cover: The table below may be helpful in determining which chapters to go through in using this book in a course.

Intended Use of the Text	Suggested Chapters
Main text for a two-semester credit course	1–3, at least two others, plus either Chapter 9 or 10
Main text for a three-semester credit course	Most of the book
Supplementary text	1–3 plus at least one other chapter

Regardless of how you use this book, Chapter 1 is essential reading. It sets up how to think about mathematics in a cultural setting.

Here are a few questions that may come up regarding teaching a course in cultural mathematics (and using this book):

- *What if a student starts describing mathematics of a culture I know nothing about?* This kind of situation is something to look forward to. Again, no one, including you the instructor, is expected to be an expert on everything. So, when someone begins describing a culture you do not know about, it is a great opportunity for that student to learn more about that particular culture. In other words, instead of having you bear the responsibility of learning about the culture and explaining it to the class, the students should be the ones who do the learning and explaining. What I have done in the past is the following:
    - Include some unspecified activities in your course syllabus where students have some choice of topics. That way, you can respond to a

student's comments on a culture (or topic) that you are not familiar with by saying something like, "That sounds very interesting. For one of the essays (or class presentations), you can choose that as your topic." In my experience, a short essay (1–3 pages) or an individual or group class presentation has worked well.

  - It is important to have accountability in the above point. You should require that students include a minimal number of formal sources (i.e., published books or journal articles) in their essay or presentation. The bibliography in this book has a fairly large list of sources with which I am familiar, but if you want to make sure students access as many sources as possible, it might be worth your while to either talk with a librarian at your school or arrange for that person to go through the process of looking for resources (either in your class or at the library).

- *What if a student asks a mathematical question I can't answer?* This is another good "teaching moment." It could easily happen that a curious student asks a question for which the mathematical answer is quite difficult to explain. For example, in the chapters on games and calendars, there are questions that can come up whose explanations quickly go beyond the level intended in this book. So, what to do? First, if you feel the topic is definitely beyond the scope of your course, you should probably just say so. However, do not end the discussion there. You can refer the student to the bibliography in this book where there are some mathematical references or you can refer the student to a mathematician whom you think could explain the concepts to the student. If you know a mathematician whose expertise could be useful in your class, you may consider inviting her or him to your class to explain some mathematical details.
- *Won't students think the course material is "watered down" because memorizing formulas and theorems are not emphasized?* My answer is, I hope not! In my experience I have found that it is best to maintain an activity ratio of about one third between the short answer, calculations, and essay/discussion activities. Students who have strong backgrounds in calculating activities may find such activities easier; however, they will still have to participate in short answer (connecting concepts), and discussion/essay activities, both of which require distinct types of thinking.
- *What if I fall so far behind schedule that I will not be able to cover all topics?* In every course I teach a strive for quality over quantity. If you are behind your schedule, you can start choosing the topics you consider most crucial to the goals of your course. In most chapters the early discussions and examples tend to be less time-consuming, and you can cover such parts earlier and plan to come back later for more depth if time permits. At any rate, you should not feel obligated to race through the material just to say you finished it. Another point to consider is that this

> book contains somewhat more (though hopefully not excessively so) material than what one would cover in a typical course. The reason for this is to give instructors some flexibility in how to run their own courses. For example, if you want to spend a lot of time discussing calendars, then you will find plenty of material in Chapter 8 on that topic. Emphasizing that chapter will probably mean you will have less time for other chapters, but that flexibility is up to you.

This work grew out of my experience in cultural mathematics. Some of that experience includes courses in ethnomathematics that I have taught at the University of North Dakota and at the Instituto Tecnológico Autónomo de México (ITAM), and from giving presentations on topics within cultural mathematics at several universities and conferences. The rest of my experience comes from about 12 years of exciting experiences in learning and researching various aspects of cultural mathematics. I have been inspired by a few specific works, namely, Ascher (1991, 2002), Zaslavsky (1999), and Closs (1986). As a reader, you will notice the influence of those references in this book. Finally, although I have tried to include examples of cultures from as many parts of the world as possible, there is somewhat more material about cultures in the Americas. This is because most of my experience is with cultures in this geographical area.

I am grateful to those people who have reviewed and commented on various parts of the original manuscript. In addition, I am thankful to many colleagues and acquaintances I have made in places such as Mexico and Peru who have been very helpful to me. In particular, I have learned a great deal about Otomi culture from people directly or indirectly involved with the Seminario de Cultura Otopame at the Anthropological Institute at the National University of Mexico (UNAM). Thanks also go out to Leonor E. Alcántara G., who has been an excellent source of information concerning anthropological concepts and references. I also thank many students who have been in my cultural mathematics/ethnomathematics courses and have inspired me with their interesting responses and questions about the topic. I am very thankful to Susanne Steitz-Filler at John Wiley & Sons for editing and help in getting this work published. Last, but by no means least, I would like to thank my family for their support during this project.

All diagrams, drawings, and photographs that are not cited from a specific source were made by me.

THOMAS E. GILSDORF

Grand Forks, North Dakota
December 2011

PART I

GENERAL CONCEPTS

1

UNDERSTANDING THE CULTURE IN MATHEMATICS

Motivational Questions: Before you begin reading this chapter, take some time to think about your responses to the following questions.

- *What is culture?*
- *What is mathematics?*
- *What kinds of things constitute mathematical ideas or activities, apart from equations, theorems, and so on?*
- *Given two distinct cultures, in what ways are they different? In what ways are they the same?*
- *How do people from a given culture communicate?*
- *Is it possible for people from distinct cultures to understand and explain the same mathematical concept differently? If so, can you think of any examples of this?*
- *What kind of mathematics would you find in reading or watching the news?*
- *How important is accuracy to you? Do all quantities have to be accurately measured? If not, explain an example of a quantity that is typically not accurately measured.*
- *Do you know exactly how many pens and pencils you own (or are in your household)? Why or why not?*

Introduction to Cultural Mathematics: With Case Studies in the Otomies and Incas, First Edition.
Thomas E. Gilsdorf.

- *How important is writing to you? Is it possible to learn and/or remember a significant amount of mathematics without writing?*

INTRODUCTION

Welcome to this book! You probably opened it with an interest in the relationships between culture and mathematics. In this book you will find that there are many fascinating connections between mathematics and culture; the Table of Contents will give you a rough idea of what to expect. The most important first step is to describe what cultural mathematics is. This means we will have to understand the meaning of "culture" and "mathematics." First, let us contemplate:

What is culture?

You may wish to think this over or discuss it with others before continuing on to the next paragraph.

This book is about mathematical ideas that have been developed by people in numerous cultures, so we will describe what is meant by the term "culture." Perhaps you will be surprised to know that the definition of the term "culture" is not so simple. This is because there can be subcultures within cultures, and there are distinct cultures that share common activities. So, let's get to the idea of a culture, the essence of which is contained in this: *As human beings we tend to assign meanings and beliefs to events in our universe*. Events in our universe that affect us most profoundly include births, deaths, courtship, child rearing, how and what we eat, how we construct our shelters, how we exchange goods with others, and, of course, how we speak (language). We as humans typically assign some meanings or beliefs to these and other events that occur in our lives. *When a collection of people follow a similar trend in assigning meanings and beliefs, they have what anthropologists call a* ***culture***. For our purposes, we will identify specific cultures based on how they have been identified by cultural anthropologists. That is, if a particular group of people is considered by anthropologists to form a culture, then we will too. See chapter 1 of Robbins (2009) for more details on the definition of culture. I hope this gives you a beginning idea of what a culture is.

However, there is an issue of how we interpret the actions of people who are from cultures other than our own. Our interpretation of other cultures is dependent on our own view of the world, and people from a particular culture that we are discussing may see themselves quite differently from how we view them. For example, have you ever eaten grasshoppers or ants? In some parts of the world, eating insects is common. In other parts of the world, such as in the United States, eating insects is almost unheard-of. Physically speaking, humans can eat many species of insects without ill effects, so the fact that some cultures choose to include them in their diets, while others do not, is a cultural, not physical choice. This is an example of a difference between cultures. In our context of mathematics, we will see that the ways in which a culture interprets mathematics can also be very different from how mathematics is interpreted in other cultures.

The interaction between culture and mathematics is the topic of this book. Having described an idea of culture, we continue with another fundamental, yet not so easy, question:

What is mathematics?

You may wish to think about this or talk it over (with a friend, in your class, etc.) before continuing.

For many people, to describe "mathematics" might be to cite some specific activities such as solving equations, using variables, proving theorems, and so on. These are certainly parts of mathematics, but what about activities such as weaving a shawl, learning about your kinship relations, playing games, or cooking? Is there anything mathematical about these? Let's look at a couple of examples and see what conclusions can be made. The first is about cooking.

Example 1.1 Suppose we want to make æbleskivers, a traditional Danish pancake. The following is a recipe.

Ingredients: 2 eggs, 1 cup flour, 2 tsp. baking powder, 1/4 tsp. salt, 1 cup milk, powdered sugar and butter (or margarine) as needed. Apple slices, cinnamon: optional.

Instructions: Separate eggs and beat egg whites stiff. Sift dry ingredients. Add egg yolks and milk to dry ingredients and beat until smooth. Fold in egg whites. Put about 1/2 teaspoon butter or margarine in the bottom of each indentation of a pre-heated æbleskiver pan. Fill openings with batter. Insert a slice of apple in each opening if desired. Turn when the batter gets bubbly, using a knitting needle or chopstick. The æbleskivers should be spherical in shape. Keep turning until the knitting needle or chopstick comes out clean from the center. Sprinkle the æbleskivers with powdered sugar. Serve with jam (lingonberries, for example), or syrup. Makes about 18 æbleskivers.

If we are going to make æbleskivers, our first step will be to read the *explanation* of the process of making æbleskivers. For instance, we will have to understand what the quantities of a cup and and fractions of a cup are; we will have to understand the quantity of teaspoons and fractions of teaspoons; and we will have to know what an æbleskiver pan is (it is a heavy cast iron pan with several circular indentations in it, each one having a diameter of about 2 inches). We will have to *measure* some quantities. We will use the concept of *number* regarding the values of the measurements, number of eggs, number of æbleskivers at the end, and so on. Specifically, it turns out that the number of indentations in an æbleskiver pan (at least in the one I own) is seven. The recipe states that we will have about 18 æbleskivers when we are done, so this will mean filling the æbleskiver pan twice, then filling four openings of the pan after that. If we want to make a different quantity, then we will have to adjust the quantities in the recipe accordingly.

On mixing the ingredients, we transform them into something called "batter," and as such, we are *designing* the batter starting from the ingredients. Finally, when we fill the holes of the æbleskiver pan we are using the concepts of "part versus whole," and we identify the "shape" of a sphere. We could say that we *locate* the shape of a sphere as the shape formed by filling the holes of the æbleskiver pan and heating.

Well enough. Now let us consider another example.

Example 1.2 What are the dimensions of a rectangular room with area 216 square feet such that one side is 6 feet longer than the other?

To solve this problem, we first note that it a *measurement*, in this case, of a room. Next, we will use the concept of *number* by stating that the unknown x will designate the length of one side of the room, as a number in units of feet. We *locate* the geometric object called a "rectangle" and an equation that describes the area of a rectangle: $(length)(width)$. Our next step is to *design* an equation that describes our particular situation. It looks like this: $x \cdot (x + 6) = 216$, and then we can rewrite it as: $x \cdot (x + 6) = x^2 + 6x = 216$. We finish the problem by constructing an *explanation* of the process of finding the value of x: For the sake of conversation, let us choose the technique of completing the square to solve it. We take one half of the coefficient of x, square it, then add that square to both sides of the equation:

$$x^2 + 6x + 9 = 216 + 9, \text{ so } x^2 + 6x + 9 = 225, \text{ and } (x + 3)^2 = 225.$$

Taking square roots of both sides, we get $x + 3 = \pm\sqrt{225} = \pm 15$. Then $x = -3 + 15 = 12$, and $x = -3 - 15 = -18$. For practicality purposes, we will choose the positive value of $x = 12$, and note that the other length is $x + 6 = 12 + 6 = 18$. So, the dimensions are 12 by 18.

A realization: Making æbleskivers uses some of the same thinking processes as solving a problem from algebra! Of many other possible observations we could make about these two processes, the two I would like to concentrate on are the following.

- Making æbleskivers, although at first does not appear mathematical, turns out to be a process that does indeed involve mathematical thinking.
- If we asked both people to describe their activities, they would give us two very different descriptions despite the fact that each understands similar mathematical ideas.

The main points of this book are the above two observations, in which the context will be culture. They appear as follows:

- Cultural activities that at first do not appear to be mathematical often involve the use of mathematical concepts and thinking.
- People from distinct cultures can be knowledgeable about the same mathematical concepts, yet express them and interpret them in completely different ways.

We will see the above two themes occurring many times. It will turn out that culture-specific activities, such as weaving and embroidering, kinship relations, games, and calendars, frequently involve mathematical thinking and concepts that we can identify. What should be clear from the two examples is that rather than state a one-line definition of mathematics, we will have to describe "mathematics" in a general way, a way that includes both of the above examples within its description. That is our next task.

DESCRIBING MATHEMATICAL IDEAS: BISHOP'S SIX

In this section we will answer the question of "what is mathematics," discovering that, in fact, that answer is *a description of activities that we can identify as mathematical in content.*

If we encountered some people who were cooking and asked them what they are doing, do you think they would reply with "Oh, we are doing mathematics"? Probably not. This is interesting to us as learners of cultural mathematics because it indicates that different people interpret and understand mathematics differently. Although there is more than one way to describe what is meant by "mathematical ideas" (see, e.g., Wood, 2000, and the introduction of Ascher, 2002), I have chosen to follow the thinking of Bishop (1988). Bishop created six general categories of activities that describe what we would consider as mathematics. Below are short descriptions of each of these six categories.

Counting: This is usually the first and easiest of mathematical activities to identify, sometimes denoted as *number*. Counting exists in virtually every culture. We will see in Chapter 2 that the interpretations of counting and of numbers can vary greatly from culture to culture.

Locating: This category has to do with symbolizing the environment, which we could think of as the interpretation and representation of spatial structure. This category includes geometry; however, it is much more general than that. It includes conceptualizations of space, which Bishop describes with three levels: physical or object space, sociogeographic space, and cosmological space. Pinxten et al. (1983)

created a list of many terms that are part of locating. A partial list is shown below:

Near, separate, continuous, part/whole, bordering, bounding, overlapping, internal/external, converging/diverging, dimension (one, two, three), preceding/following, deep (depth), left/right, geometric notions, map, scale, navigating, finite/infinite, coordinate systems, resting/moving, open/closed, horizontal, vertical, upright, wide, broad, bounded/unbounded, cardinal points and cardinal directions, linear, pointing, parallel or being at an angle, surface or volume in sociogeographical space, resting/moving, (being on a) path or oriented, having direction of movement, absolute/relative, continuous/discontinuous, homogeneous/ heterogeneous.

It is important to realize that perceptions of spatial orientation vary between cultures. For example, the Navajo perceive space as dynamic, as opposed to perceiving space as static, as in Western culture. See Bishop (1988: 29–30) and Ascher (1991: chapter 5) for more details.

Measuring: The category of measuring includes comparing, ordering, and quantifying qualities. Typical examples include distances, volumes, weights, and areas. On the other hand, not all cultures emphasize the accuracy of such measuring in the same way (I will elaborate on this in the next section).

Designing: This category is about how humans create objects from the physical, social, or intellectual universe. In many cases it has to do with making things that are used for home life, trade, adornment, games, war, and religion. On the other hand, designing includes creating objects that are not physical, such as the creation of a kinship system (see Chapter 4), or the creation of an equation as we saw in the algebra example.

Playing: We may think of the word "playing" as describing a kind of game or leisure activity. However, "playing" in the context of cultural mathematics has to do with activities involving strategy, probability, and other aspects that we will consider in Chapters 6 and 7. Also, "playing" can include certain activities that have important religious or social significance.

Explaining: This is a general term that describes how humans create abstractions and formalisms. It includes aspects such as similarity (i.e., how distinct objects are in some way similar) and classifying. From the æbleskiver example, we could create an abstract concept of "batter" that consists of a thick liquid used for making a variety of foods. The abstract concept of a variable is another example from the aspect of explaining.

The above description will be our basic set of guidelines for identifying mathematics. Our next objective is to determine what is involved in the study of cultural mathematics.

CULTURE AND MATHEMATICS

We have described something about what culture is and what mathematics is. Now it is time to put the two together. Here is the definition that we will use: *Cultural mathematics is the study of activities that we can identify as mathematical, from one or more cultures*. You may have heard similar expressions for what this book is about, other than *cultural mathematics*. Other names that include the same kinds of topics as what we will consider include: *ethnomathematics*, *multicultural mathematics*, *traditional mathematics, indigenous mathematics*, and *oral and written mathematics*.

Our goal is to understand the mathematical ideas of another culture in a way that is similar to how the people of that culture understand those same ideas. This sounds simple enough, but we will see in the rest of this chapter that there are several factors connected with culture that make understanding mathematical ideas of another culture a subtle and not-so-obvious process.

We hope to understand a culture's mathematics in the same way as the people in that culture do; however, we must use our own understanding of culture and mathematics in order to do that. Thus, it will take a substantial effort on our part to try to keep our eyes and minds open when considering the mathematics of a particular culture. In fact, many cultures, including most of the cultures we will see in this book, do not even consider mathematics as a separate topic!

Here is an example of how we must try to step back from our own culture in order to understand the mathematical thinking of other cultures. Try reading this:

> *.to used are we what from different order an in languages their read cultures many ,Yet .order different a in words read to us to strange seems It .right to left from reading to accustomed are We*

The words all look familiar, but it is written in a way that is very different from what we are used to. Try reading it starting from the last word "We" and going from right to left. You can now make sense of it, but notice that reading in this different order is much slower than what we are accustomed to, and it is harder to get the meaning of what is being said. Our first impression of this kind of communication might be to see it as less efficient than our way of communicating. Yet people who read with this kind of structure everyday get to be as good at it as we are with our own way of reading. At first it may seem unnatural or strange to us, but as we learn the new way it eventually begins to make sense and we can understand how the communication system works. The same situation holds for mathematics. Many cultures understand and express mathematics in ways that are not familiar to us. It takes some time and effort on our part to get used to viewing mathematics in a new way, and the goal is to understand the connection between the mathematics and the culture that created it. Next, we will look at how cultures are motivated to think about mathematics.

A natural question regarding culture and mathematics is to consider why a culture develops mathematics. It would seem that having some information on why people develop mathematics could give us some insights. There are three aspects to consider.

- In many cases, mathematics is developed in response to a specific need (or perceived need): environmental, cultural, or social.
- Not every need leads to a development of mathematics, and cultures are not obligated to use mathematics to address every perceived need.
- Mathematical development can occur without being connected to any particular need of a culture.

Some elaboration of these points may help. The first one is commonly cited as the motivation for developing mathematics. For example, a group of people living in an arid location may develop mathematical ideas necessary to construct an irrigation system that solves the problem of obtaining water. Solving a problem based on the physical environment is certainly a natural (pun intended) setting for developing mathematics; however, the context can also be cultural or social. Most cultures have developed rules for how people are considered related, that is, kinship rules. Kinship is an example of a cultural need in which mathematical development occurs. We will encounter the mathematics of kinship in Chapter 4.

On the other hand, the second item above indicates that developing mathematics does not always have to be part of a solution. In the example of people living in an arid location, instead of deciding to develop an irrigation system, they could decide to move somewhere else where there is a direct access to water. In such a case mathematics is neither developed nor used to solve the problem. In this case, mathematical activities in the culture are taking place, but in order to discover them, we may need to look at other aspects of the culture apart from environmental needs. For example, many Native American cultures have a strong emphasis on adapting to the natural environment instead of seeking to control it. This makes understanding the mathematics of such cultures a subtle process. In the end, we do find interesting mathematics in cultures that do not have the same outlook on the world as we do. See Closs (1986) and Norton-Smith (2004) for more details on mathematical ideas of Native American cultures.

Finally, the third item points to mathematics being developed without any "reason." Indeed, most Western mathematicians would say that they often think about mathematics simply "for the fun of it"! In this context, mathematics is not being employed to solve a particular problem or need. Compare this to reading or playing a game, in which the objective is simply enjoyment for the sake of leisure and not with a specific goal in mind. Mathematical activities occur in this setting also. Now, our next task will be to understand how our own cultural outlooks affect our perceptions of mathematics.

CULTURAL TENDENCIES AND PERCEPTIONS OF MATHEMATICS

Every child grows up being part of at least one culture. Culture comes to us through our parents and relatives, the food we eat, the music we listen to, the stories that are told to us, social and religious events in our community, and the friends we have, as well as from many other sources. Because each of us experiences culture on a daily basis during our lifetime, cultural tendencies and perspectives can be simultaneously strong and subtle. Imagine, for instance, that you and another person put on tinted eyeglasses. Yours are tinted yellow and the other person's are tinted green. If each of you looks around a room and describes the colors of various objects, then, of course, each person will have a different perception of the colors of the objects. Now, if you can imagine wearing the tinted eyeglasses everyday for 10 years, your perceptions of colors would eventually be quite strong, and were you to then replace the yellow tinted eyeglasses with the green ones, many colors would seem strange to you. The long-term subtle effects of interpreting things in a particular way is what we seek to understand in this section.

If our goal is to learn about something (e.g., mathematics) of a culture that we have not grown up with (or do not have deep experiences with), then the first step for us is to recognize the influences of our own culture and how those influences can affect our perceptions of other cultures. As I alluded to in the previous section, trying to understand a concept from another culture represents a paradox. On one hand, if we are to understand how the other culture perceives the concept, we must describe it with terms and expressions that are familiar to us. On the other hand, using our own terms and expressions to describe how another culture perceives a concept is inherently inaccurate, because the people of the other culture would not describe the concept the same way we do. The point is that as we encounter activities of other cultures, our own perceptions tend to get in the way of understanding how the other culture perceives the activity. Because I have written this book in English, its readers will typically have deep experiences with Western culture (such as that of the United States, Canada, or Australia). Thus my focus here will be on how to recognize cultural tendencies of Western culture, at least when it comes to mathematics. Of course, there are variations of interpretations of mathematics (or anything else) by each individual, even among Western mathematicians.

Nevertheless, there are some views of mathematics that occur frequently enough to warrant discussion. Also, keep in mind that the point here is not whether these tendencies are good or bad, but rather that we need to be aware of them as we study the mathematics of other cultures. As you read this book, it will be important for you to try to be conscious of these views. The following is a list of characteristics that I believe describe the *Western view of mathematics*.

- A strong emphasis on the importance of number and numerical accuracy.
- A tendency to assume that other cultures need to or want to follow a path of development similar to how Western culture has developed. In particular, a tendency to assume that the mathematics of every culture is constantly "evolving" into mathematics that eventually will be similar to Western mathematics.
- A tendency to emphasize writing and to place less importance on accomplishments by cultures that do not have an identifiable system of writing.
- An assumption about gender roles; in particular, that work done by women does not include any mathematics.

It is worthwhile to consider some examples of the above. Let us start with the emphasis on number and accuracy. The news media are a good example: in many cases topics are described using numerical information. News about economics and related topics is often stated in terms of numbers of points that a stock exchange has increased or decreased, perhaps even restated as percentages. Information about foreign markets or corporations is also described with numbers that indicate, for example, how many people are unemployed or how many of some product have been sold recently. Similar information can be found in sports, weather, and even in the general news arena.

To see how the above represents a cultural emphasis on number, we can ask ourselves questions such as: Is knowing the specific numerical information crucial to understanding the news reports? Is knowing the numerical information necessary for daily survival? The general answer to both of these questions is "no." In fact, you may wish to try the following experiment. Have someone you know read the news for several days, then ask them to repeat some specific numerical information such as stock exchange values over the past several days (do not indicate in advance which information you will ask about). You will probably find that the other person cannot repeat much of the numerical information. Why? Because knowing the details of the numerical information is not the most important part of the news. Suppose that, instead, the economic news report described situations by using phrases such as, "Economic activity has improved over the past week." This may seem strange to you to hear news this way, but knowing the general situation is probably enough for you to understand how things are going. Introducing large quantities of numerical information is not likely to give you a substantially different image of the current economic situation. Naturally, we can observe that some people, such as professional economists, do feel a need to know numerical information in great detail. Also, you may feel distrust toward a news service that describes all news items in such vague terms. On the other hand, numerical information is often manipulated so as to give false impressions, so more numerical information does not necessarily imply more honest news! Nonetheless, for most people, having a reasonable understanding of a news report usually does not specifically require knowing a lot of numerical details of that report. The fact that Western news reports include such detail is an indication of a cultural preference. In some cultures, numbers are not considered to be crucially important, and describing events with few numerical details is quite common.

The emphasis on numerical accuracy is similar to the emphasis on number. There is a tendency in Western culture to seek accurate values of quantities, measurements, and so on. People in other cultures do not necessarily view a more accurate value as being preferable to other considerations. An example from Bishop (1979), quoted in Bishop (1988: 38–39) illustrates this point. For many people in Papua New Guinea who have gardens that are roughly rectangular in shape, whenever there is a dispute over areas, the procedure is to measure the length plus the width. For purposes of comparison, this procedure is perfectly legitimate, even though it does not yield an accurate measurement of area. When we discuss calendars in Chapter 8, we will see that it is possible to have a useful calendar even when that calendar does not depend on accurate astronomical information.

Let us now look at the second item, the tendency to assume that other cultures need or want to follow the same path of development as has occurred in Western culture. This tendency is a result of how we try to understand new situations. We tend to compare such situations with what we are familiar with, and if something has worked well in our circumstance, we are tempted to think that it would work well in the new situation that we have encountered. In the case of mathematics, cultures have created ways of interpreting and expressing it that are very different from what we are accustomed to. Our goal is to understand how the *other* culture perceives mathematics.

Here is an example of the assumption of other cultures following the same path of development as in Western culture.

Example 1.3 Referring to the type of words used for counting, Smith (1923: 8), states, "the Niuès of the Southern Pacific use 'one fruit, two fruits, three fruits,' and the Javans use 'one grain, two grains, three grains,' all these being relics of the concrete stage of counting."

The assumption being made in this quote is that all cultures go through a "concrete stage of counting" and eventually move on to a more abstract form of counting, as has occurred in Western culture. However, there is no reason to believe that a culture must necessarily develop what we consider more abstract forms of mathematics—in this context, counting—in order to be successful. We will see examples throughout this book of cultures that understand nontrivial mathematical concepts without having developed the same kind of mathematics as has been developed in Western culture.

Another example I would like to mention has to do with writing, the third item in the list of Western views of mathematics. There is a tendency in Western culture to emphasize writing in the development of a culture. Nevertheless, over the course of human history, the number of cultures that have developed a system of writing is quite small. In Chapter 10 we will look at the Inca culture of South America and see that they did not develop what in Western culture would be considered a system of writing, yet they did develop sophisticated mathematics. Below is an example.

Example 1.4 Consider the following quote from Eves (1990: 11) regarding number symbols of early societies: "And still later, with the refinement of writing, an assortment of symbols was devised to stand for these numbers."

We can see that there is an assumption being made that in order for mathematics to be developed beyond the stage of simple counting, it is necessary to have a system of writing. Yet we will come across many examples of rather complex mathematical understanding that occurs outside the context of writing. We can better appreciate the mathematics of other cultures by realizing that mathematics can be created in many ways, apart from using writing.

The fourth tendency has to do with the assumption that tasks done mainly by women do not include any mathematics. We have already seen an example of cooking, an activity often attributed to women, that involves mathematics, though you probably would not find cooking discussed in typical mathematics texts. In Chapter 5, on art and decoration, we will see that textile art, an activity often associated with women, also includes nontrivial mathematical ideas in its process. Another example is the following.

Example 1.5 Consider the following quote by H. Bent regarding the book *Crocheting Adventures with Hyperbolic Planes*, by Daina Taimina, which was awarded the "World's Oddest Book Title" for 2009: "On the one hand you have the typically feminine, gentle and woolly world of needlework and, on the other, the exciting but incredibly unwoolly world of hyperbolic geometry and negative curvature" (Higgens, 2010).

You can see that an assumption is being made that the activity of crocheting, indicated as a task done by women, should not deserve to be connected with a mathematical topic (in this case hyperbolic geometry).

As you read through this book, it will be important to be able to recognize the Western views of mathematics described previously. Let us look at one more example. More examples for you to contemplate are in the exercises.

Example 1.6 In the following quote, describe which of the Western views of mathematics has been assumed: "When the numbers of tally marks became too unwieldy to visualize, primitive people arranged them in easily recognizable groups such as groups of five . . ." (Burton, 2007: 2).

Observe that what is being described is a situation in which "primitive" people felt a need to start grouping numbers into larger blocks, in this case into blocks of fives. It is an assumption that all cultures want to "progress" toward using larger bases for counting. But in fact many cultures did not follow this path. We will see in the next chapter that the Siriona culture of what is now Brazil did not develop a complicated system of counting (see Closs, 1986: 16–17). Yet the Siriona do understand many aspects of mathematics.

ISN'T MATHEMATICS "CULTURELESS"?

Some people have stated that mathematics is a universal language, independent of culture (see, e.g., Darling, "Mathematics as a Universal Language"). A property such as the commutative property of sums of numbers (that is, $a + b = b + a$) is true no matter how the numbers are described. It would seem that there is no cultural aspect to mathematics. So, how can a book such as this one on cultural aspects of mathematics even exist? To answer this question, we can first observe that properties that are universally true are by no means restricted to mathematics. The ability of humans to perceive our physical world is the same for all cultures. That is, colors, smells, textures, sounds, and so on are received by the bodies of all humans in essentially the same way. If several people stick their hands into a large bowl of mud, each will sense the same properties of texture and temperature of the mud, but each person would describe the sensations differently. As *cultural* creatures, recall that our tendency is to assign meanings and beliefs to our universe. This assigning of meanings and beliefs is what distinguishes how different people perceive the same physical world. Now, to see how culture really does play a role in mathematics, let us use an analogy from music.

What Western musicians call the note "middle C" occurs at the same sound frequency regardless of which instrument or voice is used to create it. Yet few people would call music "cultureless." People and cultures assign meanings and beliefs in the form of *interpretations*, *expressions*, and *emphases* of tones such as "middle C." These occur in numerous, creative ways with a wide variety of instruments, voices, and combinations with other tones (which would be called "notes" as well). Moreover, specific combinations of sounds or uses of particular instruments that are emphasized by some people or cultures are not given the same importance by others. In this way, there are cultural or personal *emphases*. All of this creativity and cultural influence occurs despite the fact that tones like "middle C" are "the same everywhere."

Example 1.7 Describe the cultural aspects of the *right triangle property*, more commonly known as the *Pythagorean Theorem.*

The property expresses the fact that the square of the length of the hypotenuse of a right triangle equals the sum of the squares of the lengths of the other two sides, or as you were probably required to memorize once upon a time: $a^2 + b^2 = c^2$ (where c represents the length of the hypotenuse, while a and b represent the lengths of the other two sides). This property is true regardless of how one describes the lengths of the sides and the squares of those lengths; hence it appears that there are no cultural aspects to this result. What gives the property a cultural aspect is the way it is given meanings and beliefs, that is, the ways in which it is *interpreted*, *expressed*, and *emphasized* in different cultures. For example, in Western culture the property is expressed as a symbolic equation: $a^2 + b^2 = c^2$. Moreover, the property is strongly emphasized as being an important example of abstract geometric properties. Other cultures need not express the property the same way, nor is it necessary for other cultures to emphasize the property as being important the way it is in Western culture. In fact, we can ask the same kind of question of the Pythagorean Theorem that we asked about the importance of number. It would go something like this: "Is it necessary for a culture to emphasize the understanding of this property in order to develop mathematics?" The answer is, in general, "no." For example, it is widely known that the Maya culture of parts of Mexico and Central America developed mathematics that is considered quite sophisticated by Western mathematicians (see Closs, 1986: 291, and Chapter 8 of this book). However, there is no indication that the Maya ever used the right triangle property. Perhaps the Mayans knew the result; but in any case, they did not emphasize it the way it is emphasized in Western culture.

WHEN CULTURES INTERACT

Interactions between cultures are inevitable, and when two or more cultures come in contact changes occur that affect many things, including mathematics. It is an unfortunate fact of human history that many instances of contact between cultures have been a consequence of one culture attempting to impose itself onto another. The result is that there is frequently terrible violence, oppression, slavery, genocide, and, in general, a deliberate and systematic attempt to destroy the other culture. There are countless historical events of this nature that have occurred since the beginning of human activity, taking place in all parts of the planet. A discussion of details of particular events and cultures that have been involved in such conflicts is beyond the scope of this book. On the other hand, it is not beyond the scope of this book to point out that, being a learner of cultural mathematics, you will find that as you delve deeper into the mathematics of a particular culture, you will find it necessary to take into account conflicts and contacts that have occurred between the culture you are studying and the other relevant cultures. You may well find contradictory information, people who are skeptical about discussing past events even if they relate to mathematics, and so on.

Of course, contact between cultures is not always so negative. Nevertheless, keeping our focus on mathematics, it turns out that even when there is a sincere attempt by one culture to understand another culture's mathematics, the resulting descriptions often contain obstacles to a true understanding. The following lists the main types of such complications.

- The persons describing the information may be unaware of their own cultural tendencies and perceptions of mathematics.
- The description could have been obtained many years ago (perhaps hundreds of years ago), when ideas about what constitutes concepts such as culture and mathematics were very different from the way they are now.
- The persons describing the information were not necessarily anthropologists or mathematicians, and their motives for creating the descriptions were not necessarily for the purpose of objectively understanding the other culture.

We will find examples of the above complications in many parts of our journey through cultural mathematics. One quick example of such complications can be seen in Beeler (1986: 111–112), in which one finds that in the 1700s, Spanish missionaries in what is now California routinely assumed and insisted on counting by tens while they were in contact with the Chumash culture. It turns out that the Chumash number system is based on 20, not 10. This misinterpretation of the Chumash counting system has led to confusion regarding Chumash mathematics. In Chapters 9 and 10, about the Otomies and Incas, respectively, there will be several aspects of their mathematics that are difficult for us to describe fully because of the types of complications arising from cultural interaction, as described above.

Another aspect of contact between cultures that affects mathematics is the fact that as two or more cultures remain in contact for long periods of time, cultural tendencies between the cultures begin to merge and the distinction between them becomes more difficult to determine. On one hand, this makes it difficult to say with confidence which concepts and activities (in our context, mathematical ones) were original to each culture. On the other hand, as two cultures merge, a new culture is created that may contain interesting combinations of concepts and activities from each of the cultures. In Chapter 2, on number systems, we will encounter examples of number expressions from particular cultures where it is not so easy to determine the origin of the expression. The difficulty comes about precisely because we are unable to determine whether the expression came from the culture at hand or was obtained by that culture through its interaction with another culture. Beeler (1986) presents more details on how mathematical expressions merge when cultures have come in contact.

I hope you have found this chapter interesting. A couple of final comments are in order. Classroom mathematics is usually presented without any context. The topic of cultural mathematics is one in which the context of the mathematics is in fact the most important consideration. I hope that after reading this book, you will see that mathematics from any culture or context, including mathematics that you are familiar with in your life, is created by *people*. The beauty of cultural mathematics is that it is really a study of human creativity.

EXERCISES

Short Answer

1.1 Is there anything mathematical about reading the newspaper? If so, give some examples of this.

1.2 Is there anything mathematical about how you get to school or work? If so, describe the details.

1.3 Why is studying cultural mathematics a paradox?

For Exercises 1.4–1.9 below, describe which of the Western views of mathematics is or are being implied. Explain your reasoning.

1.4 From Smith (1923):

- **(a)** Describing the decimal system: "There is no way to know when the world adopted this scale, but its wide geographic range leads to the belief that it must have been before the migrations [of people]" (p. 10).
- **(b)** Describing Hypatia, considered the first known female in Western mathematics: "Hypatia was the first woman who took any noteworthy position in mathematics, and on this account, . . . she has occupied an unduly exalted place in history" (p. 137).

1.5 From Calinger (1999: xv): "This search for precision, which includes ever closer approximation techniques, is characteristic of mathematics."

1.6 From Burton (2007: 1), "Our remote ancestors . . . must have felt the need enumerate their livestock, tally objects for barter, or mark the passage of days."

1.7 From Yam (2009): "Mitch Gordon, director of school relations for Reach to Teach, said his organization has seen more than a 100 percent increase in applications in the last six months, with 3,784 applicants compared to 1,488 during the same six-month period last year. The application system doesn't track U.S. applicants separately, but Gordon estimates more than 70 percent are from the United States."

1.8 From Eves (1990: 14): "We have already mentioned the use of marks and notches as early ways of recording numbers. In such devices, we probably have the first attempt at writing. At any rate, various written number systems gradually evolved from these primitive efforts to make permanent number records."

1.9 From Rosenthal (2009): "Each morning, about 450 students travel along 17 school bus routes to 10 elementary schools in this lakeside city at the southern tip of Lake Como. There are zero school buses."

1.10 If you buy 2 pounds of grapes, how many grapes is that? What does this say about the importance of number in Western culture? You may wish to repeat this question for a bag of rice, or a package of blueberries.

Calculations

1.11 Suppose you want to plan a garden and have the following information about four types of plants you will include in that garden. Plant A grows well when it is next to plant C but not next to plant B. Plant B grows well next to either plant C or plant D, but not next to plant A. Plant C grows well next to either plant A or B, but not next to plant D, and plant D also grows well next to either plant A or B, but not next to plant C. You also have the following information from the instructions on planting: Each plant needs a row of a minimum of 20 feet, and there must be a space of at least 1 foot between rows, including 1 foot of space along the edges of the garden.

(a) Describe as many of Bishop's six mathematical activities that are involved in setting up a plan for your garden as you can, by examining how activities such as counting, locating, measuring, designing, playing, and explaining are used in the plan. For locating, include some of the descriptions of space (near, separate, etc.; see the section on "Bishop's Six" above).

(b) Now set up the details of the garden plan, including a diagram of the planned garden that shows the information you have included such as the size (area in square feet) of the garden and which plants are next to which others. Keep in mind that there is more than one way to set up the garden.

1.12 Suppose you are going to install ceramic tile in a shower stall, with the following information. The stall is 3½ feet deep, 3 feet across, and 7 feet high. Also, the tiling process starts at a bottom corner and any tiles that must be cut should be in the corners. There are two soap dishes of dimensions 6 inches by 6 inches that will be put on two of the walls of the stall. The tiles you will use are all 12 inches by 12 inches (1 square foot) each.

(a) Describe as many of Bishop's six mathematical activities that are involved in setting up a plan for your tiling project as you can.

(b) Now set up the details of the tiling plan, including a diagram of the stall that shows where the soap dishes will be, where you will start the tiling process, where the cut tiles will go, and how many tiles you will need. Keep in mind that there is more than one way to set up the project.

Essay and Discussion

1.13 *Mathematics of your world.* Interview someone important in your life (e.g., a family member, colleague, close friend) about the mathematics that person does. For example, the context could be work, cultural activities, or leisure activities. Try to find out as much as you can about how the person uses the mathematical activities that we have discussed in this chapter. You may wish to use Bishop's six categories as a guide. Write, discuss, or present a summary of your interview.

1.14 Write about/discuss what the word "mathematics" means. Describe examples of activities that include mathematical ideas but that might not initially be considered mathematical (preferably different from the ones discussed in the chapter and outlined in the previous exercises). You may wish to look up definitions of "mathematics" in books on the history of (Western) mathematics.

1.15 Write about/discuss what the word "communication" means. You may wish to look up this word in a dictionary or book on communication. Describe examples of communication that are neither written nor spoken. Compare your information with how mathematics is described, for example, mathematics that is not in the form of an equation of theorem.

1.16 Repeat Exercise 1.15 for the word "culture."

1.17 Write about/discuss Western views of mathematics with respect to other cultures or societies. Describe the views. Find an example of each view and explain how the example represents the view you have claimed. Look in places such as news sources, books on history, or sources on the history of mathematics or science.

1.18 Write about/discuss the role of money in Western culture. Is there an emphasis on money? Compare the importance of money in Western culture with how it is perceived in other cultures and describe how this could affect the understanding of mathematics for people from various cultures.

1.19 Repeat Exercise 1.18 above for the concept of time.

1.20 Choose one of the following activities and write about/discuss the mathematical activities (Bishop's six) that are involved in that activity. State the objective of the task. Be specific about which tasks involve which mathematical activities.

- **(a)** Organizing a closet, in which there are adjustable shelves, and not all objects would fit on the same size shelf.
- **(b)** Sewing a garment from a cloth.
- **(c)** Learning to play a musical instrument. You may wish to specify the instrument.
- **(d)** Learning a foreign language.
- **(e)** Learning a new sport. Note that Bishop's category of *playing* applies here, but consider aspects such as strategy, and don't forget to consider other aspects of Bishop's six in addition to playing.

1.21 After reviewing the section on how mathematics is not "cultureless," write about/discuss your reaction to this: "Art is cultureless."

1.22 Investigate, then write about/discuss the origin of the dollar sign ("$"), including an explanation of how this sign represents contact between cultures.

1.23 Repeat Exercise 1.22 above for the term "algorithm."

Additional Comments

- The discussion of cultural interpretations of mathematics draws us briefly into the interesting world of cultural anthropology. There is much more that can be said about cultural tendencies, how cultures change and interact, and so on. The main reference I have used and would recommend for the interested reader is Robbins (2009).
- Reference to the word "algebra" allows us an excellent opportunity to point out an example of how mathematical ideas move from one culture to another, in this case from the Arab/Muslim cultures to Western culture. The word "algebra" comes from a book written by Mohammed ibn Mûsâ al- Khowârizmî (ca. 780–850), *Hisâb al-jabr w'al muqâbalah*. When his work made its way to Europe, the corruption of the *al-jabr* part of the title of his book eventually became the word "algebra." See Burton (2007: chapter 5) for more information.

- My discussion here of Western views of mathematics is not meant to be exhaustive, merely to point out the most obvious ways in which mathematics is often viewed in Western culture. If you read through modern sources in anthropology that discuss Western cultures, you will find more detailed information about Western views of such concepts as money, time, and other aspects of culture. One example is in Robbins (2009), where you can find discussions of money (pp. 86–102) and time (p. 129).
- The subtle nature of cultural perspectives cannot be overstated. It is safe to say that it is not possible for someone from one culture to describe aspects of another culture (which the person does not have deep experiences with) in a way that is completely objective. I grew up in Western culture, and although I try my best to avoid such subtle Western perspectives of other cultures in this book, inevitably there are places where they blur the true nature of the cultural context (hopefully only slightly). As another example of this subtle interference, see Norton-Smith (2004) for a careful analysis of Denny's article (1986) on Ojibwe and Inuit mathematics. Although Denny's overall treatment of cultures such as the Ojibwe and Inuit is positive, the presence of Western views of mathematics is present and quite subtle.
- At one point I stated that each person belongs to at least one culture. In reality, most people also belong to a second subculture or culture. For example, Western mathematicians have specific ways of communicating ideas (especially mathematical ones), and there are certain formal ceremonial-type activities such as honoring an influential mathematician with a conference when that person turns 60 or 70 years old. Anthropologists would say that with these particular activities, being a Western mathematician amounts to belonging to a separate culture. Other examples of cultures and subcultures would include people from particular parts of a country where specific foods are eaten and certain activities are considered important, in such a way that the foods and activities are not as common in other parts of the country. We might call them "regional differences," but in many ways they are actually cultural differences.
- My description of Bishop's six mathematical activities (Bishop, 1988: chapter 2) does not follow his original description verbatim. There are two reasons for this. One is that the context of his description is an analysis of mathematics education from a cultural perspective. My context is that of describing mathematics from a cultural perspective but not directly with regard to education. The second reason is simply that I have interpreted Bishop's description from my own understanding of what I think mathematics is. Although I have adhered closely to his description, there are some minor differences. For example, I have not emphasized the aspect of mathematics being created in order to satisfy a particular need of a culture as much as he does in his work. As you now know after reading this chapter, different people can and do have different interpretations of mathematics!

- One point that is particularly relevant in the context of thinking about mathematics from an anthropological view is the fact that, in general, anthropologists have abandoned the use of the word "primitive" to describe cultures that have not followed the same path of development as in Western culture. The motivation for not using this term is to avoid implicit implications that non-Western cultures are necessarily inferior to Western culture. Yet the use of the term "primitive" still persists in many modern texts on the history of Western mathematics.
- For another interesting book about textile arts (as a task often assumed to be done by women) and mathematics, see Belcastro and Yackel (2011).
- The study of mathematics and science within the context of culture is an area that is receiving significant attention. Two recent examples are Katz (2007) and Selin (2008).
- Throughout the rest of this book I will use CE for "Common Era" in place of AD, and BCE for "Before Common Era" in place of BC.

2

NUMERATION SYSTEMS

Motivational Questions: Before you begin reading this chapter, take some time to think about your responses to the following questions.

- *How many ways can people count?*
- *What kinds of mathematical structures do words like "fourteen," "fifteen," and "thirty-three" represent?*
- *What base does your numeration system use? Is it possible to count using more than one base? Is it possible to count without any base at all?*
- *How important are numbers and counting in Western culture? How important are numbers and counting in your daily life?*
- *Is it possible for a culture to place almost no importance on counting? What would such a situation mean to the daily lives of people in that culture?*

INTRODUCTION

In what kind of ways can we count? Computers "count" by using only zeros and ones. In the United States, elementary school children learn to count up to numbers like 100 by learning how to count by fives, by tens, by twenty-fives,

Introduction to Cultural Mathematics: With Case Studies in the Otomies and Incas, First Edition.
Thomas E. Gilsdorf.

and, of course, by ones. We also encounter the use of Roman numerals, such as in the preface of this book, or on the dates of older movies. These examples already give us an idea that there is more than one way to think about counting. People in different cultures around the world count or have counted in a wide variety of ways. In this chapter I hope to show you some general ideas of counting and some specific examples of how people in other cultures count. The main ideas are about the counting base, which describes the way numbers are formed in the language of the culture, and the mathematical properties that can occur in counting. In the next chapter, we continue the discussion by looking at how numbers are represented as symbols. In this chapter and the next, the mathematical activity we are considering is, of course, counting.

COUNTING BASES AND MATHEMATICAL PROPERTIES OF NUMERATION SYSTEMS

In order to give it a formal name, let us agree to refer to the way a particular culture or group of people counts as its *numeration system*. I will refer to a word or phrase that represents a particular number as a *number expression*. A *base* of a numeration system is an integer that represents how blocks of numbers are counted. In modern English counting is done using the decimal, or base ten, system. For example, the expression "fifty-seven" indicates a process of five tens followed by adding seven. Of course, different cultures may express this same value in different ways. The value "fifty-seven" might be expressed as a process of taking two twenties, then adding ten, then five, then two; that is, "$57 = 2 \times 20 + 10 + 5 + 2$." In this case, a combination of terms using fives, tens, and twenties is being used. Thus, there can be more than one base. We will see examples of cultures that count in combinations of two or even three bases, and there are cultures that have only a few number expressions and do not use a specific base for counting. Throughout the history of humans, people have counted using many different bases. The most common bases are five, ten, and twenty. Do you know why? Think about fingers and toes.

There is also the question of *mathematical properties* involved in a numeration system. In English we have number expressions like "fifteen," which can be described as "five" and "ten," that is, "five plus ten." Notice that the expression is not "five plus ten," but rather "fifteen." The point is that although neither of the words "five" nor "ten" appears in the expression, the similarity of "fif" to "five" and "teen" to "ten" shows us a pattern of adding five to ten to express the value. The expression for fifteen represents an addition. In the Opata language of the northwestern part of what is now Mexico, the word for nine is *kimakoi*, which can be described using the two parts: *ki* and *makoi* (ten). It turns out (see also Chapter 9) that this word in fact describes nine as "ten minus one." This is an example of the use of subtraction. There are many creative combinations for expressing numbers including addition (the most common), subtraction, multiplication, and even a few cases of division. In this chapter we will look at some specific examples of languages that use a variety of mathematical structures in their number expressions.

When we encounter the number expressions from other cultures, we can try to determine what base or combination of bases (if any) is being used, and which mathematical properties are used to construct numerical expressions. We will see that such a determination is not always so easy, and there are many cases in which we cannot say with complete certainty what the structure of the number expression is. Sometimes this is because the expression represents a new, independent way of describing the number that is not connected with other number expressions. The word "eleven" in English, for instance, is not "oneteen," and hence does not appear to be connected with any other number expression. In other situations the structure is embedded in the expression but is not obvious. If we think again about "fourteen," the word "four" is evident, but we deduce that "teen" really represents "ten," and there is an inherent assumption of addition, even though nothing like "add" appears in the expression. When determination of the mathematical structure of a number expression gets complicated, we have to appeal to the discipline of linguistics. The help of a linguistic anthropologist may even be needed in order to analyze number expressions and determine their possible mathematical structure.

NUMERATION SYSTEM EXAMPLE

Let us look at some specific examples. We start by analyzing the mathematical structure of the Assiniboine, a cultural group of the northern plains of North America.

Example 2.1 The number expressions of the Assiniboine culture.

Table 2.1 shows a list of some Assiniboine number expressions. Take a few minutes to study them and search for patterns.

Now, our goal is to search for mathematical properties that are being used in the formation of the number expressions. We also hope to be able to state what the *base* of the Assiniboine number system is. First, take a close look at the words for the values of one, two, three, and up to ten. There are a few common letters being used in some of those words, but little else. Our initial guess would be that these number words are probably not related to each other. It is fairly common that the first several number expressions of a culture are unrelated, so this observation is not surprising. Once we get past ten, a pattern begins to emerge. Notice that the expression for eleven contains the word *wą̬zį̬́*, which means *one*. It is not yet clear what *aké* means. So, let us go

TABLE 2.1 Assiniboine Number Expressions (from Barriga Puente, 1998, with original reference of Levin, 1964)

Number	Assiniboine Expression	Number	Assiniboine Expression
1	*wązį́*	13	*aké yámni*
2	*nų́pa*	14	*aké tópa*
3	*yámni*	20	*wikcémna nų́pa*
4	*tópa*	21	*wikcémna nų́p sam wązį́*
5	*zápta*	22	*wikcémna nų́p sam nųp*
6	*šákpe*	26	*wikcémna nų́p sam šákpe*
7	*šakówi*	30	*wikcémna yamni*
8	*šaknóḡa*	40	*wikcémna tópa*
9	*napcíyąka*	80	*wikcémna šaknóḡa*
10	*wikcémna*	100	*opáwiḡe*
11	*aké wązį́*	1000	*koktópawiḡe*
12	*aké nų́pa*	1,000,000	*wóyawa tą́ka*

on to the expression for twelve, which contains the word *nų́pa*, or *two*. Next, the expression for thirteen contains the word *yámni*, which means *three*. Thus, it appears that the word *aké* must mean something about adding to ten. Our first conclusion could be that the Assiniboine culture counts using base ten; however, it is best to look further down the list to see if other bases might be in use. The expression for twenty is *wikcémna nų́pa*, which looks like "ten" and (i.e., multiplied by) "two," and the expression for twenty-one, *wikcémna nų́p sam wązį́*, looks like "ten times two plus (*sam*) one." Other expressions (you can find more in the references for Table 2.1) up to one hundred show a pretty clear indication of the use of base ten (exclusively) in Assiniboine counting. Finally, notice that the word for 100 is a new expression. The expression for 1000 appears to be something like "ten hundreds," and that for 1,000,000 is again a new expression. Although this is only a partial list of number expressions and there may be some expressions that do not fit the pattern, we can conclude that for the most part, the Assiniboine numeration system is a decimal system. If the study of the Assiniboine number system were part of a detailed investigation of Assiniboine mathematics, we would want to check more examples of their numeration.

In our next example we look at the mathematical structure of the Yoruba culture of western Africa.

Example 2.2 The number expressions of the Yoruba.

The Yoruba culture is from West Africa, mainly Nigeria, but also Benin. Details of the Yoruba people and their mathematics can be found in Zaslavsky

TABLE 2.2 Yoruba Number Expressions (from Verran, 2000)

Number	Yoruba Expression	Number	Yoruba Expression
1	*kan*	11	*mọ́kònlaa*
2	*méjì*	12	*méjìlàá*
3	*mẹ́ta*	13	*mẹ́talàá*
4	*mẹ́rin*	14	*mẹ́rìnlàá*
5	*márùún*	15	*mẹ́ẹ̀ẹ́dogún*
6	*méfà*	16	*mẹ́rìndínlógún*
7	*méje*	17	*mẹ́tadínlógún*
8	*méjo*	18	*méjìdínlógún*
9	*mẹ́s'an*	19	*mọ́kọ̀ndínlógún*
10	*mẹ́wa*	20	*ogún*

(1999) and Verran (2000). Table 2.2 shows a list of Yoruba number expressions.

As we did with the Assiniboine, we seek mathematical structures in the Yoruba expressions. Looking over the words for numbers one through ten, there are some similar-looking beginnings to several of the words, but they do not appear to be connected. In the expression for eleven, we see that it does not seem to fit a pattern, either. However, the expression for twelve, *méjìlàá*, contains the word for the number *two*, which seems to indicate adding two to ten. Also, the expression for thirteen, *mẹ́talàá*, contains the word for *three*. This starts to look like base ten counting as we saw with the Assiniboine. Indeed, the expression for fourteen follows the pattern. Now for fifteen: *mẹ́ẹ̀ẹ́dogún*. Wait a minute, the pattern does not follow! There is nothing that looks like expressions for *five* and only *mé* from *ten*. Now what? We can look further down or up the list for clues. The word for twenty is *ogún*, and this suffix appears in the expression for fifteen. Maybe fifteen has to do with *subtracting* from twenty. Let's check sixteen: *mẹ́rìndínlógún*. It also contains a suffix similar to *ogún*, but even more telling is that it contains a prefix that resembles the number *four*. The pattern is looking more like a subtraction from twenty; that is, that fifteen is five subtracted from twenty, sixteen is four subtracted from twenty, and so on up to eighteen: *méjìdínlógún*. For nineteen, *mọ́kọ̀ndínlógún*, it is not so clear if this is a subtraction of one from twenty; however, the prefix *mọ́kọ̀n* appears, as it did in the word for eleven. We can guess that *mọ́kọ̀n* has something to do with *one*.

We seem to making progress, but we are mainly *guessing* the structures of the number expressions. It is very important that our conclusions be solid, both for our own dignity and as a gesture of respect for the culture we are considering. We must try to avoid a Western tendency of assuming that a different culture sees mathematics the same way we do. What we would need to do to proceed at this point is to learn more about the Yoruba language. In general this could mean finding out details of the language as described by linguists, anthropologists, or historians. It may be worthwhile to find a native speaker of the language, with a translator if necessary. In the case of Yoruba numeration, we are fortunate that Helen Verran's article (2000) is quite detailed in its descriptions.

So, let's make use of her article and look closer at the Yoruba numeration system. Starting with the expression for twelve, *méjìlàá*, we see the expression *méjì* for *two*, but now we must learn what *làá* means. Verran explains it this way (2000: 349): "The verb *ó lé* (it adds) is used in various forms. For example, numerals eleven through fourteen incorporate the element *láà*, a contraction of *ó la éwá*, as in *méjìlàá* (twelve)." We can now say with confidence that the words for twelve through fourteen represent a base ten structure. On the other hand, notice how *láà* is quite different from *ó la éwá*. It would not be easy to see this connection unless we were very familiar with the Yoruba language. In general, it can turn out to be quite difficult to determine the structure of a number expression, especially if little is known about the language in question.

For fifteen through eighteen, Verran explains (2000: ibid.): "The involvement of subtractions is indicated by the inclusion of the verb *ó dín* (it reduces) in various forms. . . . For example there is *mẹ́rìndínlógún* (twenty it reduces four—sixteen). . . ." Finally, she explains the prefix *mọ́kòn* (2000: ibid.) as another form of the verb *ó lé*, hence indicating that eleven, *mọ́kònlaa*, does indeed mean to add one to ten, and nineteen, *mọ́kọ̀ndínlógún*, means to subtract one from twenty.

Let us summarize what we know so far about the structures in Yoruba numeration. The words for one through ten are independent words. Following that, we have structures:

$$11=1+10, \quad 12=2+10, \quad 13=3+10, \quad 14=4+10, \quad 15=-5+20,$$
$$16=-4+20, \quad 17=-3+20, \quad 18=-2+20, \quad 19=-1+20,$$

and the expression for twenty is independent. Notice how subtraction is used in the Yoruba numeration. This subtraction is an example of the kinds of things we discussed in Chapter 1 about how other cultures understand and interpret concepts. Subtracting values is quite different from the English decimal structure, so at first it may seem like a strange way of counting to us. Nevertheless, people from the Yoruba culture who express numbers this way on a daily basis get to be good at it, and this way of counting eventually becomes second nature to them.

Now to another question: What *base* is being used in the Yoruba numeration system? We have seen that in some cases numbers are added to ten and in others, numbers are subtracted from twenty. One thing that will certainly help is if we look at more number expressions. Table 2.3 shows more Yoruba expressions along with the structure of each. If you look up Verran's work

TABLE 2.3 More Yoruba Number Expressions (from Verran, 2000: 347–348)

Number	Yoruba Expression	Structure	Number	Yoruba Expression	Structure
25	*mẹ́ẹ̀ẹ́dọ́gbọ̀n*	−5 + 30	50	*àádọ́ta*	−10 + (20 × 3)
26	*mẹ́rìndínlọ́gbọ̀n*	−4 + 30	51	*mọ́kọ̀nléláàád óta*	+1 − 10 + (20 × 3)
28	*méjìdínlọ́gbọ̀n*	−2 + 30	52	*méjìléláàád ọ́ta*	+2 − 10 + (20 × 3)
29	*mọ́kọ̀ndínlọ́gbọ̀n*	−1 + 30	55	*márùúndínl ọ́gọ́ta*	−5 + (20 × 3)
30	*ọgbon*	30	60	*ọgọ́ta*	20 × 3
31	*mọ́kọ̀nlélọ́gbọ́n*	+1 + 30	63	*mẹ́tàlél ọ́gọ́ta*	3 + (20 × 3)
34	*mérìnlélọ́gbọ́n*	+4 + 30	70	*àádórin*	−10 + (20 × 4)
35	*marùúndínlógójì*	−5 + (20 × 2)	74	*mẹ́rìnléláàád ọ́rin*	+4 −10 + (20 × 4)
36	*mérìndínlógójì*	−4 + (20 × 2)	80	*ọgórin*	20 × 5
37	*mẹ́tàdínlógójì*	−3 + (20 × 2)	90	*àádọ́ran*	−10 + (20 × 5)
39	*mọ́kòndínlógójì*	−1 + (20 × 2)	100	*ogọ́rùún*	20 × 4
40	*ogójì*	20 × 2	150	*àádọ́jọ*	−10 + (20 × 8)
41	*mọ́kọ̀nlógójì*	+1 + (20 × 2)	1000	*ẹgbẹ̀rùún*	200 × 5
45	*márùúndínláàád ọ́ta*	−5 − 10 + (20 × 3)	1500	*ẹ̀ẹ́dẹ́gbèj ọ*	−100 + (200 × 8)
46	*mẹ́rìndínláàád ọ́ta*	−4 − 10 + (20 × 3)	1900	*ẹ̀ẹ́dẹ́gbàá*	−100 + (200 × 10)
49	*mọ́kọ̀ndínláàád ọ́ta*	−1 − 10 + (20 × 3)			

(Verran, 2000), you will find a much more extensive table of Yoruba number expressions and a detailed explanation of the structures. Take some time to look over Table 2.3 and try to decide what the base or bases might be.

After looking over the Yoruba expressions in Tables 2.2 and 2.3, we can now state some things about the Yoruba numeration system. First, there is more than one base being used. We could say that both base ten and base twenty are being used in Yoruba numeration. Here we have an example of a number system that uses more than one base. We could even say that there is a certain amount of counting in base five, through the subtractions to get to the next multiple of ten or twenty. Thus, our final statement on bases in Yoruba numeration is that the system is based on ten and twenty, with some base five involved. The fact that more than one base is being used may seem unusual to us, but this phenomenon occurs quite often in numeration systems. Let us look at some specific examples of constructions of Yoruba values.

Example 2.3 Explain each part of the Yoruba expression for the value of 37.

The expression, given in Table 2.3, has the following format: "$-3 + (20 \times 2)$." To explain this construction, we note that *mẹ́tà* means three. The next part, *ógóji̇̀*, is a version of the expression for 40, which is given as 20×2. The middle part, *dínl*, must be the connective that indicates "add negative to," or "subtract from." Thus, the complete expression indicates "add the negative of 3 to 20×2."

Example 2.4 Determine and explain the Yoruba expression for the value of 43.

We use the expression for 41, given in Table 2.3, as a guide. Because the subtraction forms do not begin until 45, the value of 43 must be given as an addition of three to 40. Hence, our first part is *mẹ́ta*, for three. We affix the expression for three to *ógójì*, the expression for 40 ($=20 \times 2$) as done in the expression for 41. This gives us the expression *mẹ́talógójì*, which means "$3 + (20 \times 2)$."

Example 2.5 Determine and explain the Yoruba expression for the value of 53.

We use the expression for 52, given in Table 2.3, as a guide. Because the subtraction forms do not begin until 55, the value of 53 must be given as an addition of three to 50. However, notice that 50 is expressed as "$-10 + (20 \times 3)$." This means our first part is *mẹ́ta*, for three. We affix the expression for three to *àádọ́ta*, the expression for 50 ($= -10 + 20 \times 3$) using *lél*, as done in the expression for 41. This gives us the expression *mẹ́talélàádọ́ta*, which means "$3 + -10 + 20 \times 3$."

TABLE 2.4 A Few More Number Expressions

Culture or Language	Location	Number	Expression	Structure
Toba	Paraguay	6	*nivoca cacainilia*	2×3
Coahuiltecan	TX (USA)	13	*puguantzan ajti c pil co pil*	$4 \times (2 + 1) + 1$
Hare	Subarctic Canada	20	*onk'edetté*	2×10
Shambaa	Tanzania	7	*mufungate*	$10 - 3$
Apinaye	Brazil	5	*aicluto aicluto pütchic*	$2 + 2 + 1$
Maidu	CA (USA)	8	*tsoye- tsoko*	4×2

Notice how the Yoruba number system uses several mathematical properties. Subtraction was previously mentioned, and we can also see that the Yoruba system employs multiplication by twenties, and several combinations all at once, such as $+1 - 10 + (20 \times 3)$ for 51. The Yoruba number system includes many interesting mathematical properties! To finish this section, we will take a brief look at some more examples of mathematical properties from a variety of cultures.

Table 2.4 shows a few more examples of mathematical properties that appear in number systems. I have not included division examples because they are quite rare, but you can find some such examples in chapter 1 of Closs (1986).

Consider the number system of the Siriona of Bolivia. The number one is expressed as *komi*, two is *yeremo*, three is *yeremono*, and any value more than three is *etubenia* ("much") or *eata* ("many"). In this system, there is no base at all. Other examples of number systems that do not have a base can be found in Closs (1986) and Barriga Puente (1998). The question that could come to mind is why a culture would employ such a number system. The explanation comes from the cultural view of numbers and counting, which we now discuss. You may wish to look again at Chapter 1 on general cultural considerations, too.

CULTURAL ASPECTS OF NUMERATION SYSTEMS

Let us recall the cultural significance of numbers in modern Western cultures from Chapter 1. In Western culture there is a strong emphasis on numerical information. What about the Siriona of Bolivia, who only have a few number expressions? In the Siriona culture numbers and counting do not carry the same level of importance that they do in Western culture. This is an example of how perceptions of mathematics can be radically different for different cultures. At the same time, we must be careful to avoid trying to decide which type of counting system is "better." Such a comparison is an "apples versus oranges" situation in the sense that if in a particular culture the counting system is perceived to be satisfactory, then that counting system works for that culture, regardless of how other cultures count. Moreover, the fact that a culture such as the Siriona has a counting system with few items in it does not mean that, as a culture, the Siriona are somehow using "less" mathematics. Think back to Chapter 1 about how mathematical thinking manifests itself in many ways (broadly organized into six categories described by A. Bishop). The Siriona probably use mathematics in ways other than counting, such as in their artwork or creation of social or physical objects.

In fact, it is even likely that the Siriona have a keen understanding of their environment that has not been obtained in the same way as might be done with Western mathematics. To understand how this can be, consider an article written by Richard Nelson (1993). Nelson is an anthropologist of Western culture who spent significant time with Koyukon Indians of the interior of Alaska in the United States. Although the Koyukon people he worked with did not utilize typical Western methods of understanding the arctic environment where they live, Nelson's conclusion was that the Koyukon Indians have a deep understanding of their environment, an understanding that is at least as deep as the level of understanding of arctic environments that is known to Western culture. Like the difference between counting in cultures, this is an example of how two distinct cultures can understand something (in this case an arctic environment) in radically different, but nevertheless profound, ways.

Determining the mathematical structure of number expressions, as we did in the previous section, is only part of the story. The other part of numeration is how numbers are perceived in a culture, such as the importance of certain values of numbers, or even the importance of numbers in general. We have seen a difference in the importance of numbers in the above discussion of the Siriona counting system. To add to these aspects, understanding the structure and origin of such expressions can also give us clues about how a culture has changed over time. Perhaps the culture in question used one base in its numeration system and later included another base. When we discuss the Otomies in Chapter 9, we will find that there are aspects of the Otomi counting system that include base five, base ten, and base twenty. The fact that there are several bases being used could mean that the counting system has changed over time, perhaps starting out with one base, then using another later on.

Looking at numeration systems can also give us clues about how two or more cultures have interacted. Recall our discussion of how the interaction between cultures can affect mathematics from Chapter 1. As two cultures interact with each other, each one begins to borrow things from the other, and expressions such as those for numbers begin to get blurred between the cultures. This cultural interchange becomes more noticeable if the two cultures remain in contact for a long time. Such long-term contact between cultures, be it friendly or not, is common throughout the world. For example, the English spoken in the United States contains many words from the indigenous cultures of the Americas and indicates the long history of contact between the Europeans who migrated to the United States and the cultures that were already present. For instance, the word "skunk" comes from the Algonquian word "seganku." Many other words such as the names of many states are Native American in origin. Words such as "tomato" and "chocolate" are words that were borrowed from the Aztec language of Nahuatl. If we can determine that the number expressions of one culture are related to those of another culture, then that relation indicates that the two cultures may have had some contact. For a long time linguistic anthropologists have used language as an indicator of cultural interaction between two or more cultures. What is new here is the possibility to use *mathematics* as an indicator of cultural interaction. In this chapter the mathematical activity we have been considering is that of counting, via numeration systems. As you read further in this book, you will see that there are many examples of mathematical connections between two or more cultures, such as calendars and artwork.

Certain specific numbers can be very important to a culture. What about the cultural importance of specific numbers? Even though many people might say that there is no cultural significance to numbers in modern Western society, there are indeed examples of such numbers. For example, what does the number thirteen often imply? If you live in a place where there are tall buildings (at least fourteen floors high), perhaps you can visit one such building and find out if it has a thirteenth floor. You may be surprised to find out that many do not! So, in fact, the number thirteen frequently implies bad luck. What about seven? What about one million? Unlike the current gross national product of the United States, a number that relatively few people know but which represents a cultural emphasis on number, thirteen is a number whose ritual significance is known to almost everyone (at least in Western culture). Its connotation of bad luck represents a different kind of cultural emphasis, the emphasis that thirteen is an important number without regard to whether there is any practical use for it. Some ethnomathematicians say numbers like this have *ritual* or *social significance*. Many cultures have numbers that are ritually or socially significant. We have seen thirteen as an example from Western culture. The following is an example from a non-Western culture.

Example 2.6 The Ojibwe is an indigenous group of the Great Lakes and Eastern Woodlands areas of North America. Their numeration system is decimal (see Denny, 1986). In many Ojibwe customs, we find the number *four* having a special importance. For example, in the game of windigo, popular with Ojibwe children, the child chosen to play the role of the windigo is decided by an arrangement of four sticks. Another example occurs in a love charm in which the effect is to take place in four days. In funerary practices, when a deceased person passes to the hereafter, the process takes four days. These examples are taken from Closs (1986: chapter 7), in which you can find more details and more examples of the ritual use of four in Ojibwe culture.

EXERCISES

Short Answer

2.1 There are bases other than ten that are present in modern Western culture. Briefly describe two of them.

2.2 Describe an example in Western culture in which the exact number of items is *not* important.

2.3 Why can we conclude that a person from a group having only a few number expressions would be able to understand Western mathematics such as calculus?

2.4 How do the words "eleven" and "twelve" in English fit with decimal counting? What are the origins of these words? Explain briefly. You may wish to look up these words in a dictionary that describes etymology.

2.5 Does the number three have any cultural importance in Western culture? If so, explain what that importance is.

2.6 Describe at least two examples of each of the following arithmetic constructions of number systems from specific cultures:

(a) Addition; **(b)** subtraction; **(c)** multiplication.

Calculations

2.7 Construct Assiniboine expressions for the following values. Explain how each part of the expression contributes to the construction of the value.

(a) 25; **(b)** 59; **(c)** 1,001,077.

2.8 Construct Yoruba expressions for the following values. Explain how each part of the expression contributes to the construction of the value.

(a) 38; **(b)** 59; **(c)** 73; **(d)** 77; **(e)** 95; **(f)** 153.

2.9 Consider the following partial list of number expressions from the Ekoi language of Cameroon (from Zaslavsky, 1999: 48):

3, *esa*; 4, *eni*; 5, *elon*.

From this list determine the mathematical structures of these number expressions:

(a) 6, *esaresa*; **(b)** 7, *eniresa*; **(c)** 8, *enireni*; **(d)** 9, *eloneni*.

2.10 The following is a list of number words from the Chumash culture of what is now part of southern California (from Closs, 1986):

1, *paqueet*; 2, *eshcóm*; 3, *maseg*; 4, *scumú*; 5, *itipaqués*; 6, *yetishcóm*; 7, *itimaség*; 8, *malahua*; 9, *etspá*; 10, *cashcóm*; 12, *maseg scumu*; 13, *maseg-scumu canpaqueet*; 18, *eshcóm cihue scumuhúy*; 19, *paqueet cihue scu-muhúy*; 20, *scumuhúy*; 21, *scumuhúy canpaqueet*.

Determine the following (be sure to explain *how* you made your conclusion):

(a) The arithmetic construction of the number words for 5, 7, 12, and 18. Indicate the parts of the words that indicate specific values.

(b) The base, or combinations of bases, of this counting system.

2.11 Consider the group Embërä culture from Western Colombia (from Barriga Puente, 1998: 243):

1, *ab'á*; 2, *umé*; 3, *ombéa*; 4, *kimáre*; 5, *jüësumá*; 6, *jüësumá ab'*; 7, *jüësumá umé*; 8, *jüësumá ombéa*; 9, *jüësumá kimáre*; 10, *jɨwá umé*; 11, *jɨwá umé ab'á*; 12, *jɨwá umé umé*; 13, *jɨwá umé ombéa*; 14, *jɨwá umé kimáre*; 15, *jɨwá umé jüësumá*; 16, *jɨwá umé jüësumá ab'á*; 20, *embërä ab'á*; 25, *embërä ab'á jüësumá*; 30, *embërä ab'á jɨwá umé*; 33, *embërä ab'á jɨwá umé ombéa*; 40, *embërä umé*; 100, *embërä jüësumá*.

(a) For each of the numbers 1 through 10, determine how the number word is constructed. Include an explanation of each part of the expression. It may help to list the numbers vertically.

(b) Explain the term for 20. What is its significance?

(c) Construct expressions for the values stated below, including an explanation of how the value fits with the pattern of the other Embërä expressions.

19, 27, 35, 50, 80, 86, 140, 400.

(d) Determine the base, or combinations of bases, of this counting system. Explain how you obtained your conclusion.

2.12 Construct your own counting system of a base *other than* 5, 10, or 20. Build into your system specific arithmetic constructions as has been discussed in this chapter. Include at least two such constructions. Include at least one number expression that does not fit with other patterns and explain how that expression could have made its way into the counting system.

2.13 *To be done by two people*:

(a) *Person 1*: Turn around so that you cannot see what Person 2 is going to do.

Person 2: Put a number of objects in front of Person 1. There should be enough objects (at least fifteen) so that it is not possible to know how many objects there are by looking briefly at the collection. When finished, instruct Person 1 to turn around and look at the objects.

Person 1: Look at the collection of objects for about five seconds. *Do not try to count the number of objects!* After five seconds, turn back around again.

Person 2: Remove a number of objects. Remember, the number of objects removed can be zero. When finished, instruct Person 1 to turn around and look at the collection again.

Person 1: Look at the objects again, and decide whether any objects have been removed or not. *Do not try to count the number of objects!* If you think some objects have been removed, state how many objects you think were removed.

Person 2: Explain to Person 1 how many, if any, objects were removed.

Persons 1 and 2: Describe this experiment as a presentation, oral report, or essay, indicating how accurately Person 1 was able to determine, *without counting*, the number of objects removed.

(b) Repeat the experiment in part (a) several times. Be sure to include some times in which the roles of Person 1 and Person 2 are reversed. Also, be sure to include some times in which the number of objects removed is zero.

(c) Repeat the experiment in part (a), with the following new aspects:

- **(i)** The person who changes the number of objects can either remove some objects, add some objects to the collection, or leave the number of objects the same.
- **(ii)** The person who changes the number of objects rearranges the objects in addition to part (i).

Essay and Discussion

2.14 Write/discuss some of the basic ideas of number systems, such as bases and arithmetic constructions that are most common, and why. Also include explanations and examples of number systems that utilize more than one counting base. Be sure to include cultural considerations in your discussion.

2.15 Write/discuss some of the less common number systems, that is, counting systems that include few number words or that use constructions that are not common (subtraction is one such construction). Include aspects such as possible cultural reasons why some cultures have rather unusual number expressions. Indicate examples from sources outside this book, if possible

2.16 Write/discuss the modern Western number system, including information about its origin, the arithmetic constructions that are used in this system, and why. Also include cultural considerations of the Western counting system in your discussion.

Additional Comments

- My information about the Opata language is taken from Valiñas Coalla (2000).
- For all of the number expressions in this chapter, I have used the linguistic notations as they appear in the references I give with each one. These notations indicate guides to pronunciation as determined by linguists. General descriptions of these notations can be found by looking up the International Phonetic Alphabet (IPA), for example, on the Internet.
- Alternate names for the Ojibwe culture include Ojibway and Chippewa.
- When I tried the experiment described in Exercise 2.13, I used small cubes (manipulatives) that numbered between 18 and 21 pieces for each pair of people who ran the experiment. I asked the students to give me some data numbers afterward, and asked them to discuss what the experiment indicated to them regarding a person's ability to surmise quantities of objects without directly counting them.

3

NUMBER GESTURES AND NUMBER SYMBOLS

Motivational Questions: Before you begin reading this chapter, take some time to think about your responses to the following questions.

- *How many ways can numbers be represented? Write down as many representations of the number five as you can think of.*
- *What is the difference between the representations 607, 706, and 76?*
- *What does 0 (zero) stand for?*
- *Where did the symbols 0, 1, 2, 3, 4, 5, 6, 7, 8, 9 come from?*
- *What are the differences between the representations 13, "thirteen," and* 卌 卌 ||| *?*
- *If you were in a crowded place where it would be difficult for others to hear you, how could you express numbers?*

INTRODUCTION

Throughout the last chapter I tried to avoid using symbols for numbers, writing for example "thirteen" instead of "13." In this chapter we will discuss number symbols. Under the category of number symbols, I will include representing numbers in some kind of written form or by physical gestures.

Introduction to Cultural Mathematics: With Case Studies in the Otomies and Incas, First Edition.
Thomas E. Gilsdorf.

There are several motivations for using a symbol or gesture to represent a number. One is to have a written representation. Another is for making calculations. A third motivation is that using hand or body gestures to represent numbers can be useful in certain situations where other ways of representing numbers are not so easy. Finally, a fourth motivation for number symbols is that such symbols provide a form of expression of numbers that are important for ritual or social purposes. Some cultures such as the Mayans of Central America have used special symbols for numbers when they appear as part of a social or ritual context. Apart from the symbol itself, there is the issue of how the symbols are arranged. In particular, for some types of symbols the order matters, whereas in others, it does not. We will discuss the cultural implications of these possibilities as well. To begin our discussion, let us look at hand and body gestures.

REPRESENTING NUMBERS BY GESTURES

If you are in a crowded noisy place and you want to express some number without having to shout, you may hold up some fingers to express that value. Two fingers up would mean the value two, for example. This is an example of a *number gesture*. Cultures in many parts of the world use a variety of ways to express numbers using the human body. For example, in the Klamath culture of the Pacific Northwest region of North America, the term for the number six is *nadshk-shapta*, which translates to "one I have bent down," indicating a count of five with one finger on the other hand down for a total of six. The Klamath expression for eight is *ndan-ksahpta*, which translates to "three I have bent over." You can see that these terms come directly from number gestures as representations of numbers. It may be tempting to think of using bodily gestures to represent numbers as being a simple form of expression that is not useful; however, using the body to convey complex information happens all the time. Just consider people who regularly use sign language, or consider work by psychologists who have shown how complex information is conveyed through body language. With these comments set down, let us look at a specific system of representing numbers by gestures. Figure 3.1 shows the number gestures for the values of one through nine of the Arusha Maasai culture of eastern Africa.

Notice that the gestures are not simply a matter of holding up a total number of fingers. Also, the Arusha Maasai can express numbers beyond nine in the following ways. For values of eleven through twenty, the number is expressed by making a gesture for ten, followed by the corresponding value between one and nine. The value of seventeen would be expressed by the gesture for ten, then the gesture for seven. The Arusha Maasai can express number values to at least the value of 2,000 using hand gestures. Moreover, these gestures are culturally important to the Arusha Maasai in the sense that

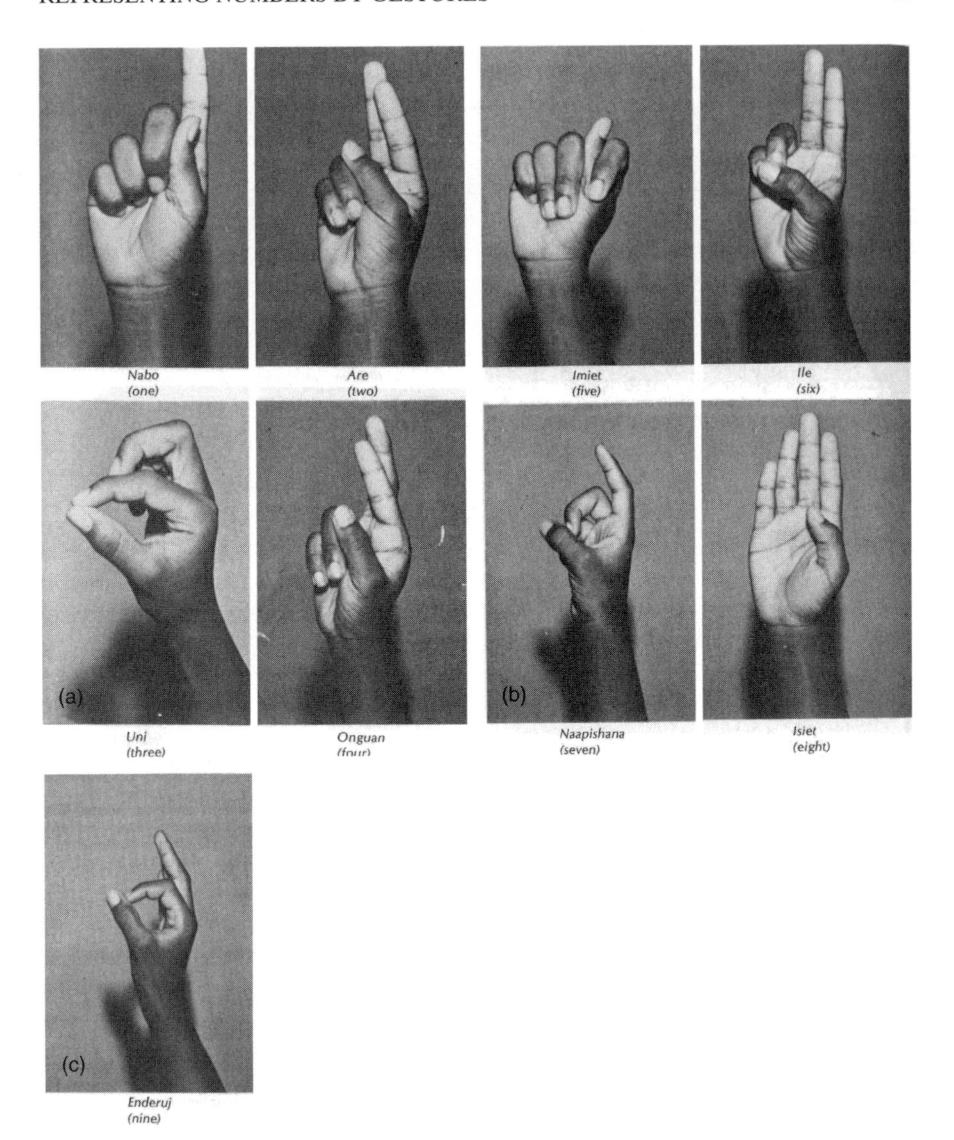

Figure 3.1 (a)–(c) Arusha Maasai number gestures for values one through nine. (Reproduced from Claudia Zaslavsky, *Africa Counts*, 3rd ed. Photos copyright: Sam Zaslavsky. With kind permission from Thomas Zaslavsky.)

during conversations involving numbers the person expressing the number is expected to use a number gesture in addition to spoken words, so as to affirm to the other person the values of the numbers being discussed. In Zaslavsky (1999), you can find more information about body gestures for numbers, particularly in African cultures. We will look at symbols representing numbers in the next section.

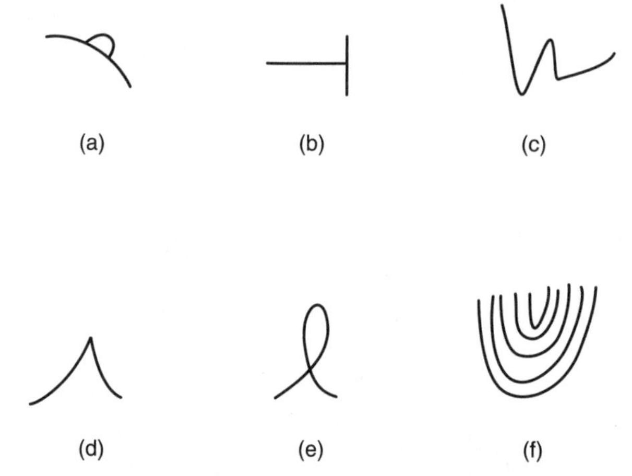

Figure 3.2 Examples of ancient number symbols from several cultures. (a) Persian symbol for five; (b) symbol from India (Nagori) symbol for 60; (c) Turkish symbol for 600; (d) East Arabic symbol for eight; (e) symbol for four originally from India and used in Germany in the Middle Ages; and (f) prehistoric marking on a petrogyph in present-day northern Mexico, a possible symbol for the number five.

NUMBER SYMBOLS

The number symbols that are probably most familiar to you are the symbols 0, 1, 2, 3, 4, 5, 6, 7, 8, and 9. These symbols originated with the Hindus and Arabs and came to Western culture between the years 1000 and 1300 (Burton, 2007: chapter 6). There are numerous number symbols that have been created by cultures around the world. Many of them are no longer in use; however, Figure 3.2 shows examples of some ancient number symbols from various cultures.

Notice the creativity involved in representing numbers! In the exercises for this chapter, and in later chapters in this book, you will encounter more number symbols, from cultures such as in China, the Maya of Central America, the Aztecs (and Otomies) of Mexico, and the Incas of Peru. We will see many examples of how useful and interesting number symbols can be. In the meantime, our next activity will be to take a closer look at a specific system of numbers symbols, namely symbols from ancient Egypt.

ANCIENT EGYPTIAN NUMBER SYMBOLS

As you already know, the modern Western number symbols listed above represent a *positional system* in the sense that changing the position of the symbols changes the interpretation of the number. For example, 73 and 37 represent different values, though with the same symbols. On the other hand, it is possible to represent numbers in ways that do not depend on position.

TABLE 3.1 Egyptian Symbols

Symbol	Value	Symbol	Value
𓏺	ones	𓂭	10,000
𓎆	10	𓆐	100,000
𓍢	100	𓁨	1,000,000
𓆼	1000	𓍲	10,000,000

Let us look at a system of number symbols that is *nonpositional*, that is, that does not depend on position. We are going to discuss a system of symbols developed by the Egyptians, starting around 3500 BCE. The symbols are in base 10. Table 3.1 shows the symbols for various quantities of powers of ten.

Example 3.1 The number 30,456,789 can be represented with Egyptian symbols like this:

𓍲𓍲𓍲 𓆐𓆐𓆐𓆐 𓂭𓂭𓂭𓂭𓂭 𓆼𓆼𓆼𓆼𓆼𓆼

𓍢𓍢𓍢𓍢𓍢𓍢𓍢 𓎆𓎆𓎆𓎆𓎆𓎆𓎆𓎆 𓏺𓏺𓏺𓏺𓏺𓏺𓏺𓏺𓏺,

or this:

𓂭𓂭𓂭𓂭𓂭 𓏺𓏺𓏺𓏺𓏺𓏺𓏺𓏺𓏺 𓆼𓆼𓆼𓆼𓆼𓆼 𓍲𓍲𓍲 𓍢𓍢𓍢𓍢𓍢𓍢𓍢

𓆐𓆐𓆐𓆐 𓎆𓎆𓎆𓎆𓎆𓎆𓎆𓎆,

or this:

𓎆𓎆 𓆼 𓏺𓏺𓏺 𓍢𓍢𓍢𓍢𓍢𓍢𓍢 𓏺𓏺𓏺𓏺

𓆼 𓏺 𓆼𓆼 𓆐𓆐 𓂭𓂭 𓆼𓆼

𓎆𓎆𓎆𓎆𓎆𓎆 𓂭𓂭 𓍲𓍲 𓂭 𓆐𓆐 𓏺 𓍲.

Notice that the order of the symbols does not matter. This system of symbols is *nonpositional*. This kind of symbol use may seem strange to us, but the Egyptians who used it were skilled at using these symbols. Moreover, we can ask ourselves: If a culture uses a nonpositional system of symbols for its numbers, would that culture be able to create all the same mathematics as in a positional system? We can investigate this by trying some mathematical operations with the Egyptian symbols and seeing what happens. Before looking at any examples, please read this: *In order to understand how the Egyptian symbols work, we will sometimes employ some Western reasoning and notation (such as "+," "−," "=," "×," "÷"). Keep in mind that the Egyptians who developed the symbols did not use these notations, and need not have expressed their understanding of the mathematical concepts (e.g., addition) as we will here.*

To add two numbers symbolically, we first must think about what happens as we collect values of numbers in a particular base, in this case ten. Each time we have a collection of ten objects, we can replace that collection with a symbol from the next higher power of the base (ten). In the positional decimal system familiar to you, the sum of 99 and 4 is represented by 103, where we have moved up from 9 powers of 10^1 to one power of 10^2. The same process applies in a nonpositional system. Let's look at an example.

Example 3.2 Addition with Egyptian symbols.

To add

we collect similar symbols from each number, obtaining

There are eleven symbols of 𓍢 and thirteen symbols of 𓆼, so we replace ten of the symbols 𓍢 with one symbol from the next higher power of ten, namely, 𓆼. Then we are left with one extra 𓍢. We replace ten of the symbols 𓆼 with one symbol 𓂭 and are left with three extra 𓆼 symbols plus the new 𓆼 generated from the eleven 𓍢, for a total of four 𓆼 symbols. Now, there were nine 𓂭 symbols, so by replacing the ten 𓆼 with one 𓂭, we now have ten 𓂭 symbols, so we replace these ten 𓂭 by one 𓆐. Our final sum is:

To subtract a second number from a first number we collect objects of the first number and then delete the value corresponding to the second number. When there is an insufficient collection of some power of ten, we borrow from the next higher power. For example, for 23 – 16 we must borrow 10 units from the 20 in 23 and put them with the 3, making an operation of 13 – 6. We can use this same process in a nonpositional system, too.

Example 3.3 Subtraction with Egyptian symbols. Let us subtract

from

Starting from the lowest terms in each number, we have two symbols [symbol] in the smaller number but no [symbol] symbols in the larger number, so we borrow from the next power of ten, in this case, [symbol]. Replacing [symbol] by 10 [symbol] symbols, subtract:

Now we move to [symbol]. In the larger number there are no more [symbol] symbols because we borrowed the only one in the last step, so we borrow from the next higher power, [symbol]. Replacing one [symbol] in the larger number by 10 [symbol] symbols, we subtract again:

To subtract the four [symbol] symbols in the smaller number from the now only one [symbol] in the larger, we borrow again. From [symbol], the next higher power of 10 is [symbol], but there are no [symbol] symbols in the larger number, so we move to the next higher power, [symbol]. We replace it by 10 [symbol] symbols, then replace one of the [symbol] symbols by 10 [symbol] symbols. This gives us:

Thus, our final form for the difference is:

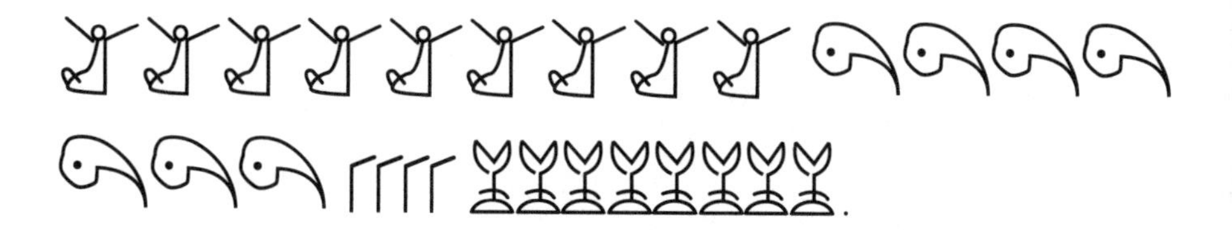

What about multiplication? To multiply two numbers, we recall that multiplication is a process of repeated addition, so one way to accomplish this would be to repeat the first number the second number of times. Let us try this.

Example 3.4 Multiplication with Egyptian symbols. Multiply
𓎆𓎆𓎆𓎆 𓎆𓎆𓎆𓎆 𓍢𓍢 by 𓎆 III II.

We take advantage of our understanding of multiplication as repeated addition, and the base, namely ten. First, eight 𓎆 multiplied by 5 I makes a total of 40 𓎆. Every 10 𓎆 is equal to one 𓍢, we have a total of four 𓍢 for this part of the multiplication. Next, notice that every time we multiply a symbol by 𓎆, we have ten such symbols which gives us one of the next level symbol. Thus, eight 𓎆 multiplied by 𓎆 makes a total of 80 𓎆, or eight 𓍢. We have a total of 12 𓍢, which gives us one 𓆼 with two 𓍢.

For the next step, two 𓍢 times five I gives a total of 10 𓍢, which equals one 𓆼. Likewise, two 𓍢 times 𓎆 is 20 𓍢, or two 𓆼. We have a three 𓆼 for this part of the multiplication.

To finish, we add all of the symbols: 𓆼 and 𓍢𓍢 from the first part, then 𓆼𓆼𓆼 from the second part. Our final result is 𓍢𓍢𓆼𓆼𓆼𓆼.

The last of the four basic operations to consider is division. Like products, we first have to think about what division of numbers means in terms of quantities.

Example 3.5 Division with Egyptian symbols.

(a) First, let us divide 𓍢𓍢𓍢 by ∩. Note that every ten ∩ gives us one 𓍢. So, 𓍢𓍢𓍢 divided by ∩ gives us ∩∩∩.

(b) Now, we will divide 𓍢𓍢𓍢 ∩∩ IIIIII by ∩∩ III. What we seek is to find out how many times we can divide the first number, 𓍢𓍢𓍢 ∩∩ IIIIII, into groups of ∩∩ III. First, let us see how many times ∩∩ III fits into 𓍢𓍢𓍢. In each 𓍢, we can fit five ∩∩, and for adding three more to ∩∩, we can write the last of those ∩∩ as ∩ II plus IIII IIII. Thus, for each 𓍢, we have four ∩∩ III and IIII IIII left over. There are three 𓍢, so we can repeat this a total of three times to get twelve ∩∩ III with a value of 24 I left over. The last 24 I is one ∩∩ III plus an extra value of I. To finish, we have the last ∩∩ III IIII which is ∩∩ III plus 4 I, so we include one more ∩∩ III in the quotient for a total of 14 ∩∩ III. Thus, our division is fourteen ∩∩ III plus a remainder of 5 I. We summarize:

𓍢𓍢𓍢 ∩∩ III IIII equals [∩∩ III] times [∩ IIII] plus II III

It is time for us to return to our question of whether mathematics as we know it in modern Western terms would be different with a nonpositional system of number symbols. We have seen that the processes of the four basic operations of addition, subtraction, multiplication, and division are conceptually the same in both types of symbol systems. What distinguishes using a positional versus a nonpositional system is how we move and arrange the symbols. So, our question boils down to asking whether the properties of real numbers are the same in a nonpositional system as they are in a positional system. Except for clarifying the role of zero, which we will discuss near the end of this chapter, the answer is "yes." Indeed, all of your favorite real number properties are independent of how the number is symbolically expressed. For instance, the property that changing the order of addition does affect the sum, known as the *commutative property*, is true regardless of whether you express symbols like this: 7 + 5 = 5 + 7, like this:

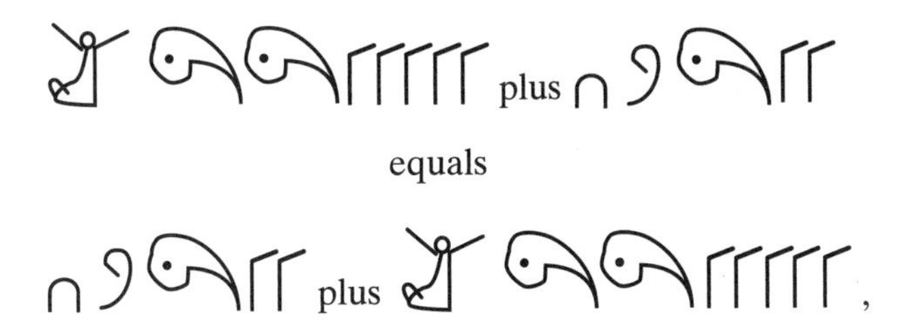

or in any other form of symbolic number expression.

If all of the properties of real numbers are true in a nonpositional system, then topics such as algebra, trigonometry, and calculus are also unchanged. Definitions of such things as decimal expansion (or base point expansion, if the base is not ten) and irrational numbers depend only on how we decide to express them with the number system we are using. Thus, the final answer to the original question is: *Yes, a culture that uses a nonpositional system of symbols for its numbers would be able to create the same mathematics as can be created using a positional system.* To illustrate, we can make use of some modern Western notation, as in the following example.

Example 3.6 Solve the equation $4x + 179 = 231$ with Egyptian symbols.

To solve an equation such as $4x + 179 = 231$, we observe that the constants 4, 179, and 231 can be represented any way we like. In effect, the equation is of the form $Ax + B = C$, so we can use whichever techniques we know of to isolate x in terms of the constants A, B, and C, then perform the necessary calculations with the specific symbols.

The solution, however we choose to determine it, comes out to be $x = \frac{C - B}{A}$. If we want to solve this using the Egyptian symbols, we calculate as follows:

A equals IIII, *B equals* 9 ∩∩∩ ∩∩∩∩ IIII IIIII, *C equals* 99 ∩∩∩ I, so
C minus B equals 99 ∩∩∩ I *minus* 9 ∩∩∩∩∩∩∩ IIII IIIII
equals 9 ∩∩∩∩∩∩∩∩∩∩∩∩∩ IIIIII IIIII
minus 9 ∩∩∩∩∩∩∩ IIIII IIII, which equals ∩∩∩∩∩ II.

We get (using a mixture of Egyptian and Western notations):

$$x = \frac{C - B}{A} = \frac{\cap\cap\cap\cap\cap\,\text{II}}{\text{IIII}} = \frac{\cap\cap\cap\cap}{\text{IIII}} + \frac{\cap\,\text{II}}{\text{IIII}} = \cap\,\text{III}.$$

Finally, if mathematics as we know it would be essentially unchanged with a nonpositional system, we can ask ourselves, Which system is better? This question brings us back to Chapter 1 because the answer depends on how a particular culture views numbers. In other words, this is a *cultural* question. The Hindu Arabic system was considered attractive partly because of a European cultural desire for speed of calculations. The Egyptian system is slower to work with mainly because some of the symbols, such as those for 1,000 and 1,000,000, must be drawn, so a value of 9,000,000 is very tedious to create with that system. Of course, when the Egyptians created their system of symbols, speed was probably not their cultural inclination. Nevertheless, had the Europeans decided to adopt a nonpositional system, speed of calculation could still have been included. For example, the symbols could have been modified so as to be faster to recreate, or ink stamps could have been made so that the figure would be stamped on paper repeatedly instead of being drawn by hand. For instance, to make the Egyptian symbols in this chapter, I created each one once on my word processor, then cut and pasted them as necessary. It was very quick compared with drawing each one every time it appeared. As mentioned at the beginning of this chapter, cultural groups such as the Mayans created special numerical symbols with the intention of portraying certain values in a ritualistic way. See Closs (1986) for more information on Mayan number symbols.

ZERO

As I alluded to earlier, some details about zero in the Egyptian number symbols must be clarified. First, if we were to ask almost anyone what "zero" means, that person would probably describe it as meaning "nothing," "no value," or something similar. However, when we write 1,002, zero takes on a different role, that of a *placeholder*. Hence, zero has two meanings in the Western number system, that of a quantity of nothing and that of a placeholder. Because of the positional nature of the system, 1,200, 1,020, and 1,002 all denote distinct values. Meanwhile, the Egyptian symbols do not include a symbol for zero, because such a symbol is not necessary in a nonpositional system. To write the value of 1,002 with the Egyptian symbols, it is 𓆼𓏻, and to write 1,020, it is 𓆼𓎆𓎆. There is no confusion about the distinction between these two values in the Egyptian system. Many cultures have developed ways of expressing zero, sometimes with a symbol and sometimes by leaving empty spaces where a value of zero could be inferred. Moreover, many cultures have expressions in their languages to express zero as a quantity. In Chapter 9 when we discuss the Incas, we will see how that culture has represented zero.

The number zero has been discussed many times in the literature. Most texts on the history of (Western) mathematics include some discussion of the "origin of zero." I have not included a description of such discussions for a couple of reasons. One reason is that many such discussions are centered on the question of which cultures developed positional systems of numerical symbols, which was the type of number system that eventually became the system in use in Western culture. Certainly, this kind of discussion is worthwhile for understanding the history of Western mathematics; however, for our context of examining all possible number systems, the concept of zero carries less emphasis. Another reason is that to discuss which cultures have understood the concept of zero, the most "indisputable" evidence of zero is that of a symbol for zero, which tends to imply that the main cultures that developed a concept of zero were the ones that also developed a writing system. Most likely, many cultures have understood the concept of zero as a value even if they did not develop a writing system. To verify such situations one would need to study numerous languages from a linguistic anthropological point of view. The idea would be to find examples of the concept of zero as expressed via language as opposed to zero expressed as a written symbol. As far as I know, no such study has been done.

EXERCISES

Short Answer

3.1 Suppose a person from the Arusha Maasai culture is conveying a number to you using hand gestures. She first makes one hand gesture, then a second gesture. What, if anything, can you say about the possible range of values being expressed?

3.2 Describe at least two situations in which using a hand gesture to express a number value would be preferable over other ways of expressing the number value.

3.3 Describe an example of a system of number symbols, still used in Western culture, that is nonpositional.

3.4 What would have been the implications in Western culture if the Hindu-Arabic symbol for zero had been very elaborate and difficult to reproduce by hand?

3.5 Explain why the following properties of real numbers are true in any nonpositional system (you may wish to look up the definitions in an algebra book): *commutative*, *associative*, *distributive*.

3.6 How would you define and represent rational numbers in a nonpositional system? How would you define and represent irrational numbers? You may wish to use the Egyptian or other nonpositional system that you know about to explain some examples of your definitions.

Calculations

3.7 (This exercise is best done by two or more persons as an activity; otherwise there will be a lot of drawing of hands.) Show the hand gestures of the Arusha Maasai for the values of one, nine, and five. Assuming a gesture for the value of ten, make hand gestures for 13, 18, and 20. Then make gestures for other values not listed here, while others in the group try to guess the value being expressed.

For Exercises 3.8–3.16, determine the sum, difference, product, or quotient, using Egyptian symbols. Do not convert to decimal-manipulate the symbols; do all operations with the original symbols, as done in the examples. (Table 3.1 is repeated here for your convenience.)

3.8 Determine the sum:

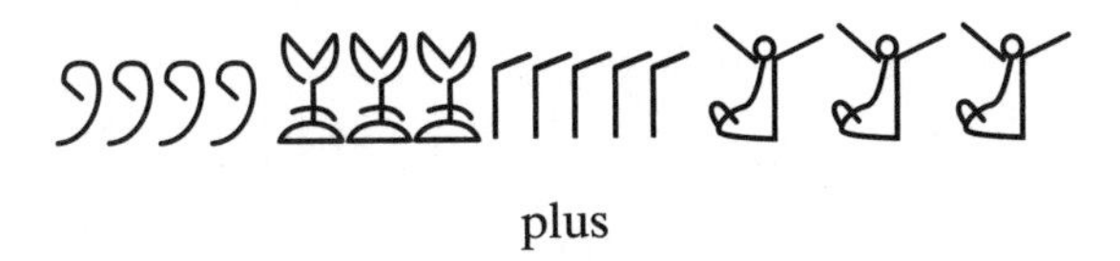

plus

𓆼𓆼𓂭𓂭𓂭𓂭𓂭𓍢𓍢𓍢𓁨𓁨𓁨𓁨𓁨𓁨𓁨𓆐.

3.9 Determine the sum:

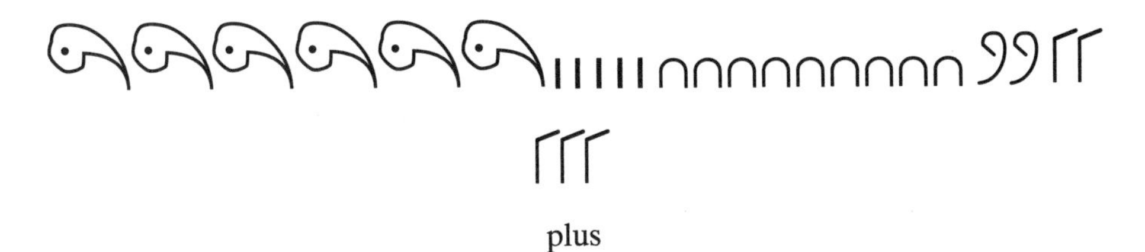

𓂭𓂭𓂭

plus

𓎆𓎆𓎆𓎆 𓆼𓆼 𓂭𓍢𓍢𓍢 𓏺𓏺𓏺𓏺𓏺 𓍶𓍶𓍶.

TABLE 3.1 Egyptian Symbols

Symbol	Value	Symbol	Value
𓏺	ones	𓂭	10,000
𓎆	10	𓆐	100,000
𓍢	100	𓁨	1,000,000
𓆼	1000	𓍶	10,000,000

3.10 Determine the sum:

𓆐𓆐𓆐𓆐𓆐𓆐𓏺𓏺𓏺𓏺𓏺𓏺𓏺𓏺𓏺 𓎆𓎆𓎆𓎆 𓎆𓎆𓎆𓎆𓎆

𓍢𓍢𓍢𓍢𓍢𓍢𓍢𓍢𓍢𓍢𓁨𓁨𓂭𓂭

plus

𓁨𓁨𓁨𓁨𓁨𓁨𓁨𓁨 𓂭𓂭𓂭𓂭𓂭𓂭𓂭𓂭𓍢𓍢𓍢𓏺𓏺.

3.11 Determine the difference:

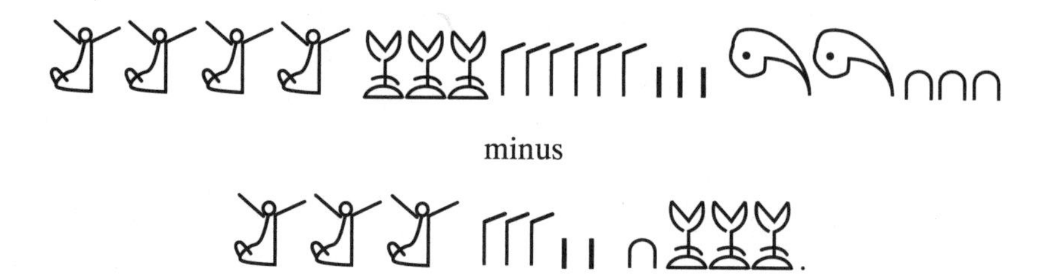

3.12 Determine the difference:

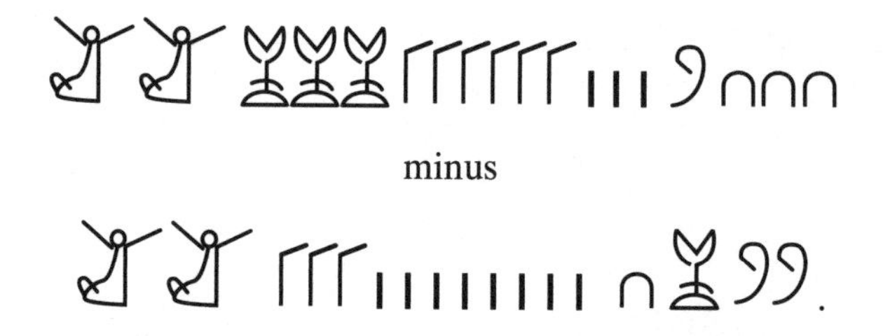

3.13 Determine the difference:

3.14 Determine the product:

3.15 Determine the product:

3.16 Determine the quotients:

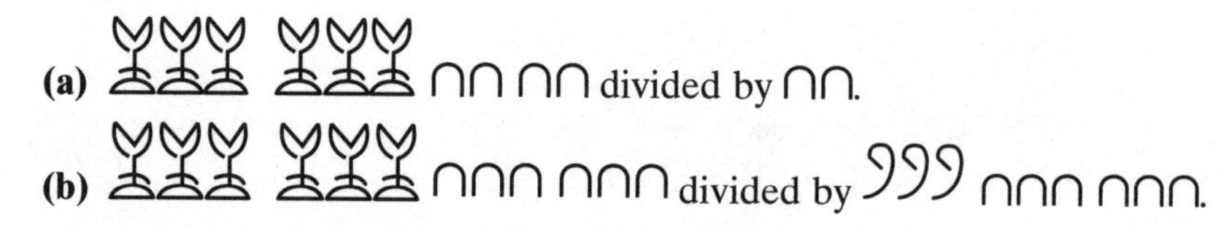

3.17 Explain how to solve the equation $3x - 5 = 13$ using Egyptian symbols.

3.18 Consider the Chinese symbols for values 1 to 10 and the powers of 10 as shown:

一 = 1, 二 = 2, 三 = 3, 四 = 4, 五 = 5, 六 = 6, 七 = 7, 八 = 8, 九 = 9,

十 = 10, 百 = 100, 千 = 1000, 萬 = 10000.

Here is an example of how these symbols work: The value of 6,437 in the decimal positional system would be expressed as:

六千四百三十七

(a) Determine the symbol expression for the number 8,762. Then explain whether this symbol system is positional or nonpositional.

(b) Determine the symbol expression for the number 3,060. Then explain whether a symbol for zero would be useful in this number system.

(c) Determine the decimal values expressed in the following symbols:
(i) 九萬八十一
(ii) 千二
(iii) 四千九百五十八萬三千一千七十七.

(d) Determine the values expressed in the following symbols, by manipulating the symbols as we did for the Egyptian number symbols. *Work with symbols; do not convert to decimal notation.*
(i) The sum of 七百八十九 and 六百九十九七.
(ii) The sum of 八萬三千四百四十七 and 六萬七千九百九.
(iii) The difference: 八萬三千四百九 minus 六萬七千九百一十九.

3.19 Table 3.2 shows one way we could create a nonpositional system that uses base seven. We can start with dots for values 1 to 6 and then arrange a number of "sticks" for each power of seven.

Thus, for 7^8, we would create a figure having 8 "sticks," and so on. As a quick example, |||▽▽▽▽•••••⌋ represents $3 \times 7 + 4 \times (7^3) + 5 + 1 \times (7^2)$ (which comes out to be 1,447). The point of this problem is to use these symbols in various calculations as we did with the Egyptian symbols.

(a) Determine the following sum using these symbols.

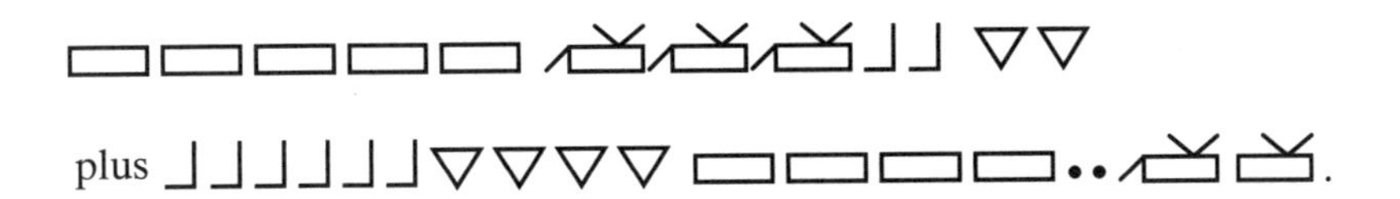

TABLE 3.2 Some Symbols for Powers of Seven

Symbol	Power of 7	Symbol	Power of 7
•	0	▭	4
|	1	⍂	5
┘	2	⍂̸	6
▽	3	∠⍂̸	7

(b) Express the difference given by the first number minus the second number from part (a).

(c) Determine the quotient: ▭▭▭▭ ┘┘ divided by ┘┘ ••▽.

3.20 Refer to the table of base seven symbols created in Exercise 3.19 above.

(a) Fill in the missing parts of the following partial multiplication table. Some of the entries are already given, but you may wish to verify them before filling in the blanks in order to get an idea of the patterns involved. The properties of real numbers (especially the commutative and associative properties) will help, too.

Times	•	••	••••	||	┘┘	▽
•				||		
••						
••••				┘|		
||					▽▽ ▽▽	
┘┘						
▽		▽▽			⍂ ⍂	⍂̸

(b) Explain some of the patterns you observed while filling in the table, and what cultural implications they may have for a cultural group that would use this system of symbols.

Essay and Discussion

3.21 Write/discuss a comparison between three kinds of expressions of numbers: number expressions (in words), body gestures, and number symbols. Indicate situations in which each would be the preferable way to express a particular value. You may wish to look up information about expressing numbers in cultures other than those presented in this chapter.

3.22 Write about or discuss what modern Western mathematics would be like if the system of number symbols were nonpositional or if a base other than ten were used.

3.23 Suppose you encounter a new culture that does not have a symbol for zero. Write about or discuss what you would do in order to investigate the number system of the culture. Describe at least two possibilities and include the conclusions that you could reach for each of those possible scenarios.

3.24 Investigate the number symbols for a culture other than those presented in this chapter. Explain what each symbol means, and show several examples of calculations with the symbols. Discuss the connection between the symbols used and the culture, for example, whether the symbols have cultural meanings.

3.25 Investigate body gestures of cultures other than those of the Arusha Maasai culture. Discuss the connections between gestures and bases for counting (Chapter 2), and give several examples of number gestures. You may wish to look at Zaslavsky (1999) and Closs (1986: chapter 1), to start.

3.26 Write/present a critical analysis of how the Egyptian symbols have been presented in this chapter. Include factors such as whether the Egyptians would have calculated with the symbols the way we did here, and, if not, then how they might have calculated such things as sums, differences, products, and quotients.

Additional Comments

- I have used the Egyptian symbols here in order to investigate the four standard mathematical operations with a nonpositional system of number symbols. The Egyptians did a lot of nontrivial mathematics, both with the symbols presented here and with others they developed over time. Most texts on the history of Western mathematics at least mention the Egyptian system. A recent, well-written work on Egyptian mathematics is Imhausen (2007), where you can learn much more about their mathematics. See also Gillings (1981).
- In calculations involving symbols, I have avoided converting the symbols to Western decimal notation. I believe that working directly with the symbols gives you, the reader, a better idea of how people of the culture at hand would have done the calculations.

- My reference for the Chinese symbols is the recent work by Dauben (2007), which I feel is very well written. You can find a much more complete discussion of Chinese mathematics in that reference and in the resources he cites.
- I have not included a discussion of how to convert between different bases using symbols in this chapter. Books about numerical mathematics or computer science often have discussions on how to change from one base to another.

4

KINSHIP AND SOCIAL RELATIONS

Motivational Questions: Before you begin reading this chapter, take some time to think about your responses to the following questions.

- *What is a relation in mathematics? Can you describe some relations between real numbers?*
- *What are kinship relations in families? Can you describe someone you know who is your second cousin, or first cousin removed? If you are familiar with a non-Western culture, how are kinship relations defined in that culture?*
- *What does the word "relation" mean in other contexts, such as in a business, a school, or the military? Can you describe how these relations are different from kinship relations?*
- *If $a = b$, then $b = a$. Is this true for the relation "$\geq$" ? What about "$<$"?*
- *How important are kinship relations with respect to issues such as property rights and inheritance?*
- *What connection is there between kinship and health practices?*

Introduction to Cultural Mathematics: With Case Studies in the Otomies and Incas, First Edition.
Thomas E. Gilsdorf.

INTRODUCTION

One of the interesting features of cultural mathematics is the fact that aspects of life that we might normally think of as being nonmathematical often contain some mathematical concepts. This chapter is about relations in a cultural sense, and the mathematics connected to those relations. First we will discuss relations in mathematics. Then we will discuss kinship relations with a focus on the Warlpiri culture of Australia. The main theme of this chapter, kinship, has important social and cultural connections. For example, in modern medical practices, it is common for a medical professional to ask questions about people related to the patient and whether they have had certain medical conditions. Also, in many cultures kinship determines such things as social status or land rights.

EQUIVALENCE RELATIONS

A great deal of effort by modern Western mathematicians is dedicated to studying relations between different kinds of mathematical objects (or concepts). Questions about how numbers or functions are related are typical in this setting. The truth of the logical statement "*If P then Q*" amounts to stating that whenever a (mathematical) property "*P*" is satisfied, then property "*Q*" must be satisfied, and hence, objects that satisfy property "*P*" and objects that satisfy property "*Q*" are related in some way. The statement "*P if and only if Q*" is a mathematical way of saying that property "*P*" objects and property "*Q*" objects are in some way "equal," and in this case property "*P*" objects and property "*Q*" objects are called *equivalent*. The objects do not have to be identical, but they have something mathematical in common. The mathematical property of equality of numbers is a good place to start in our discussion of relations.

Example 4.1 Equality of numbers. You must certainly know that the statement $a = b$ between numbers means the numbers a and b have the same values, even if they are not expressed in exactly the same way. For example,

$$\frac{12}{24} = \frac{1}{2}, \text{ or } \frac{13^2 - (8 \times 21)}{\sqrt[3]{343} - 6} = 1$$

(You may wish to check this.)

Apart from this definition of equality, the property of two numbers being equal has other features. One such feature is called the *reflexive* property, and this property represents the fact that for any number a, it is always true that $a = a$. That is, the number a is always related to itself with respect to the property of equality. What if we check the reflexive property with other mathematical relations? If the relation is “$<$,” then the reflexive property is always false because “$a < a$” is always false.

Another property is called the *symmetric* property, which for equality is manifested by the statement, *if* $a = b$, *then* $b = a$. This property holds for all numbers. It is never true for the relation “$<$“; however, for the relation “$\leq$,” the situation is not absolute. Indeed, if $a = b$, then we can say that $a \leq b$ and in return $b \leq a$. On the other hand, this symmetric property does not hold if $a \neq b$. In this case, the relation “$\leq$” is sometimes true and sometimes false.

A third property is called the *transitive* property, and it goes like this: *If a is related to b and b is related to c, then a is related to c.* For the relation of equality of numbers, this transitive property is always true. It is also true for both relations $<$ and $\leq$, as you can easily check.

If a mathematical relation satisfies all three properties, then Western mathematicians refer to such a relation as an *equivalence relation*. We have just seen that the relation of equality of numbers is an equivalence relation. In the exercises there are more examples of these properties. There are other properties of numbers we can consider, such as the commutative, associative, and distributive properties (see also Chapter 3). The bottom line is that the properties of being reflexive, symmetric, transitive, commutative, associative, and distributive are important aspects of discussions of relations between numbers. It will also turn out that such properties are important in cultural contexts. Also, there is even something to learn about relations that are not always true. Think again about “$\leq$.” In what follows we will look at some mathematical aspects of kinship relations.

KINSHIP RELATIONS

Our next topic is mathematical aspects of kinship relations. Before we start, it is important to keep in mind the general importance of kinship relations. In most cultures, relations between family members are deciding factors in such issues as defining a person's role within her or his family or community, inheritance, rights to ownership of property or land, and social status, to name a few. In the motivational questions, I asked you about your relations with family members, such as first or second cousins and first cousins removed. For people trying to trace their genealogy, these relations are particularly important. We can also consider the properties of being reflexive, symmetric, or transitive as we observed with the relations “$=$,” “$<$,” and “$\leq$” between numbers. Let's look at the relation “sibling” as an example.

Example 4.2 Consider the set of all people, and define: Person A is a *sibling* of person B if person A has the same parents as person B. Does this form an equivalence relation?

First, reflexivity: Certainly, A and A have the same parents, so A is a sibling of A. Next, if A has the same parents as B, then B has the same parents as A, so "sibling" is also symmetric. Finally, if A has the same parents as B and B has the same parents as C, then A has the same parents as C, and this means "sibling" is transitive as well. Thus, the kinship relation *sibling* is in fact an equivalence relation, like the numerical relation "=." Equivalence relations are useful for deciding categories, and in this case, "sibling" decides categories in the sense that, given any person in the world, that person either is a sibling of yours or is not. Of course, this seems pretty obvious. When we include wider relations such as "cousins," or "business colleagues," defining categories is not so crystal-clear. Even defining the relation "sister" is not so simple, as we will see next.

Example 4.3 Consider the set of all people again, and define: Person A is a *sister* of person B if person A is female and has the same parents as person B. Does this form an equivalence relation?

If we check reflexivity, we already have a problem: If person A is male, then he cannot be his own sister. So, the reflexive property does not hold. For the symmetric property there is a similar problem, because person A could be female and person B could be male. Then A is a sister of B, but B is not a sister of A. On the other hand, if person B is also female, then the reflexive and symmetric properties do hold. You can compare this relation *sister* to "≤." Recall that the symmetric property is true for "≤" only in certain cases. Finally, you can also check that *sister* satisfies the property of being transitive only when all three persons are female.

Here is another property to consider with respect to family relations: the *commutative property*. For addition of numbers the commutative property expresses the fact that $a + b = b + a$. Notice that the commutative property is also true for multiplication of numbers ($a \cdot b = b \cdot a$), but is false in general for subtraction and division (e.g., $a - b \neq b - a$). We can consider two relations such as *mother of* (denote by m) and *father of* (denote by f) in this context. Certainly, mf is not the same as fm, because the mother of the father is not the same as the father of the mother. So, in this case forming the relation of mother or father is *not* commutative. With these thoughts in mind, we begin the study of the cultural group called the Warlpiri, an indigenous culture of Australia.

Warlpiri Kinship Relations

The Warlpiri culture includes a lot of close connections with cosmology in the sense that a person is intimately connected to cycles that have transpired regarding that person's ancestors. As Marcia Ascher states regarding native Australian cultures (Ascher, 1991: 70), "The land boundaries of the tribes, the animals and plants that were to be sacred to each group, the sites that would

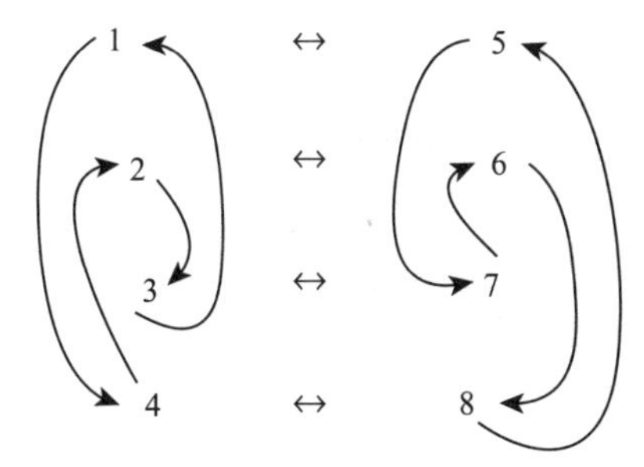

Figure 4.1 The Warlpiri kinship diagram.

be remembered in ceremonies and myths, all relate to the journey of the ancestors." Thus, kinship plays an important role in Warlpiri culture.

The Warlpiri kinship system consists of eight sections that I will describe in a moment, and each member of the culture belongs to one of the eight sections. There are social rules that determine from which section a person's spouse can be selected and to which section a couple's children will belong. Figure 4.1 shows a diagram of how the sections are related. Keep in mind that the diagram is our invention via Western mathematics and our decision of section numbers is our way of indicating relationships between the eight sections.

The double arrow symbol between the two columns indicates marriage between the two indicated sections. For example, a person from section 1 marries a person from section 5, a person from section 2 marries a person from section 6, and so on. The children of a couple belong to the section that is determined by the mother, and for each section, the arrows indicate in which section the children belong. For example, a woman from section 3 marries a man from section 7 and their children belong to section 1.

In order to understand how a person is related to his or her ancestors, we can follow the path of several generations. Because a child's section is determined by the section of her or his mother, we start by following generations of women.

If a woman belongs to section 1, her mother belongs to section 3, her grandmother (on her mother's side) belongs to section 2, her great-grandmother (mother's side again) belongs to section 4, and her great-great-grandmother (mother's side) came from section 1. Hence, the generations of women form a *cycle* in the sense that any woman in one of the sections 1 to 4 has a female ancestor, four generations earlier, from the same section. We can represent this cycle as shown below:

$$1 \rightarrow 3 \rightarrow 2 \rightarrow 4 \rightarrow 1.$$

We can call this cycle a *matricycle* because it is based on a female's ancestors. The other matricycle is from the second column of Figure 4.1. You may wish to determine it and then check to see if you get this:

$$5 \to 8 \to 6 \to 7 \to 5.$$

Because there are two matricycles that are independent of each other and together they give information on the matricycles of all eight sections, anthropologists call each group of four sections (1 to 4 and 5 to 8) a *moiety*.

Now for the males. A man from section 4 marries a woman from section 8 and his son is in section 5. His son then marries a woman from section 1 and the man's grandson is then in section 4. Thus, we can call this a *patricycle* and it goes like this:

$$4 \to 5 \to 4.$$

Another patricycle is the one that starts with section 1: $1 \to 7 \to 1$. There are four patricycles in all. I have left the details of the other two to the exercises.

It is useful to us to see another example of how the Warlpiri kinship system works.

Example 4.4 A girl is in section 8. Determine the possible sections in which a great-grandfather of the girl could be.

It is helpful to create a diagram (called a *tree diagram*) that describes the relationships between the girl and her ancestors. Such a diagram is shown in Figure 4.2. The letter G represents the girl. The next row of m and f below her represents her mother and her father. The second row represents her grandparents, and the bottom row represents her great-grandparents. Thus, the four f symbols in that bottom row represent the four great-grandfathers of the girl. Our task is to determine to which section each of those great-grandfathers belong. We can start with the left most "branch" of the tree diagram, in which we have the mother of the girl, who, by Figure 4.1, is in section 6. Her mother, (i.e., the grandmother of the girl on the girl's mother's side), is in 7, and the mother of a person in 7 is in section 5. A woman in section 5 marries a man in section 1, so the first possibility, for the left-most f in the bottom row of Figure 4.2, is section 1. Now back up a row to the grandparent level. Recalling that the mother of the mother of the girl is in section 7, that woman would marry a man in section 3, by Figure 4.1. The father of a person in section 3 is in section 6, so the great-grandfather that represents the f that is second from the left in the bottom row would be in section 6. We have two more great-grandfathers to determine, by looking at the father's side of the girl's family. Referring to Figure 4.1 again, a woman in section 6 marries a man in section 2, so the girl's father is in section 2. His mother is in section 4 (Figure 4.1 again). The father of a person in section 4 is in section 5, by Figure 4.1, so the third great-grandfather (third from the left in the bottom row; see Figure 4.2) is in section 5. Finally, the father of the father of the girl is in section 8, and his father is back in section 2.

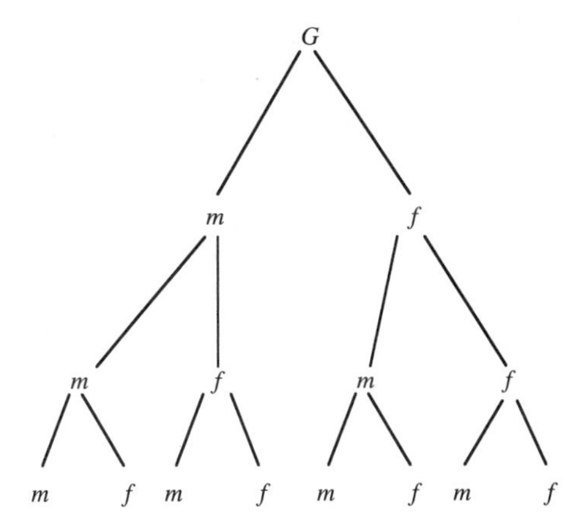

Figure 4.2 A tree diagram for Example 4.4.

To summarize, the four possible sections are: 1, 6, 5, and 2. Notice that in some of the cases of determining the sections, we could have used the fact that the father of a father is back in the same section. For example, knowing that the father of the girl is in section 2, we can say immediately that the great-grandfather of the girl, as two more generations of fathers, must be back in section 2 again.

Now let us look into the mathematical properties of the Warlpiri kinship system. As we apply the relations such as *mother* or *father*, we will refer to these mathematically as *operations*. We start with section 1, which we will denote by I. The reason for the choice of the letter I will become clear later. As with the matricycles, I will first describe the situation for females. To move through the sections starting from I, we will again use m for *mother* and f for *father*. Keep in mind that these are our own inventions and not necessarily how the Warlpiri would denote the relations. With these notations, we can say that the operation m is denoted by $m = m(I)$. Then the mother of m would be $m(m)$, which I will write as m^2. Repeating again, the next would be $m(m^2) = m^3$. After that we would have $m(m^3) = m^4$, but we know from the earlier discussion that $m^4 = I$. Referring to Figure 4.1 for the section numbers, we can rename sections 1 through 4 as follows:

$$1 \Leftrightarrow I, 2 \Leftrightarrow m^2, 3 \Leftrightarrow m, 4 \Leftrightarrow m^3.$$

We start again with section 1, that is, I, and look at the fathers. Let f denote $f(I)$. What section number corresponds to f? The mother of a person in I is in section 3, and she would marry a man from section 7. Thus, $f = 7$. Notice that this takes us to the second column of the eight sections. In order to obtain expressions for the other three section numbers, we consider mothers again because it is the mother's section that determines the section of her children,

TABLE 4.1 Warlpiri Kinship Relations Corresponding to Section Numbers

Section	Relation	Section	Relation
1	I	5	mf
2	m^2	6	m^3f
3	m	7	f
4	m^3	8	m^2f

and hence also the father's children. To this end, we calculate $m(f) = mf$, $m(mf) = m^2f$, and $m(m^2f) = m^3f$. As with sections 1 through 4, we need not calculate m^4f or higher terms because the cycle repeats. In other words, $m^4f = f$. Referring again to section numbers and Figure 4.1, we have that $f = 7$, $mf = 5$, $m^2f = 8$, and $m^3f = 6$. We have now renamed each section number as a kinship relation starting with section 1 being I. Table 4.1 shows the details.

The table certainly describes relations of the eight sections, but does it describe *all* possible combinations of m and f? We can experiment by trying a few possibilities. Before trying any example, it is worthwhile to think about the subtleties of the operations. Order for instance, is important. Indeed, we have already seen that $mf \neq fm$. In other words, unlike addition of numbers, these relations are *not commutative*.

Another property to consider is the associative property. For addition of numbers we know that this means we can add some terms first without altering the sum: $a + (b + c) = (a + b) + c$. For the Warlpiri relations, let us try to determine if the associative property is always true. In fact, we can think of this in terms of mothers and fathers because this is how relations in the Warlpiri system are determined. If your mother were to tell you about her grandmother on her father's side, that would be $mfm = mfm(I)$, where I is you. We could express this as $m(fm)$. Your grandfather on your mother's side is f. He could tell you about his mother, which we could express as $(mf)m$. Either way, it is the same person, so $m(fm) = (mf)m$. It might be tedious to do, but if we check *all* combinations of three symbols taken from the eight relations from Table 4.1, we would find that we get the same relation regardless of whether we obtain it in the form $(ab)c$ or $a(bc)$. Our conclusion is that these eight relations satisfy the associative property. We are now ready to try more examples, keeping in mind that we can group terms associatively, but we cannot change the order of the operations. Our goal is still to determine if every possible combination of relations m and f will always bring us back to one of the eight sections.

Example 4.5 Let us determine $f(mf)$.

First note that this notation really means $f(m(f(I)))$. We know that $f = f(I)$ is in section 7, so the mother of a person in section 7 is in section 5. The father of a person in section 5 is in section 4 because of the patricycle $4 \rightarrow 5 \rightarrow 4$. Section 4 is denoted by m^3, so $fmf = m^3$.

Example 4.6 Let us try this: In which section is $f^3m^{15}fm^2fm$?

We could write this out with parentheses; however, notice that in fact this and any combination is read *from right to left*. That is, we seek first m, which is $m(I)$, then fm, the father of m, and so on. Also, observe that moving from section to section through generations is connected with which section is a *mother*'s section. Here we go: m is in section 3, so that person's *mother* is in 2, who marries a man from section 6. So, fm is in section 6. The mother of a person in 6 is in 7 and the mother of that person is in 5, telling us that m^2fm is in 5. Now, the *mother* of a person in 5 is in section 8, who marries a man in section 4. Thus, so far we have that fm^2fm is in section 4. Now for the term m^{15}. We do not have to calculate 15 generations of mothers starting from section 4 because we know that each time we cycle through the relation m four times, we return to the section in which we started. In other words, as we continue determining the section of $f^3m^{15}fm^2fm$ starting from section 4 where we left off, we can use the fact that m^4 will be back in section 4 again, as will m^8. So in fact, we just need to determine how many times 4 divides into 15, and the remainder. It is: $15 = 4 \times 3 + 3$. This means we calculate m^3 starting from section 4 and that will tell us in which section m^{15} is. Starting from section 4, you can verify from Table 4.1 that three generations of mothers puts us in section 2. By the way, we could also have used the associative property to write

$$m^{15} = m^3(m^{12}) = m^3(I) = m^3$$

using the fact that m^{12} gives us I, so that we get m^3. Finally, to determine f^3 starting from section 2, we recall that all of the patricycles have only two sections in them. Mathematically, this means every time we cycle through two generations of fathers, we return to the section we started from. In other words, f^3 is in the same section as f. To finish this example, we determine the section of the father of someone in section 2: The *mother* of a person in section 2 is in section 4 and she marries a man in section 8. Our final conclusion is that $f^3m^{15}fm^2fm$ is in section 8, which we can write as $f^3m^{15}fm^2fm = m^2f$.

Now for the bigger picture. If we (tediously) check *every* combination between any two of the eight sections, we would find that such combinations always return us to one of the eight original sections. The above examples help us to see the details of how the section numbers change as we apply the various combinations. The Western mathematical term for this is called *closure* and basically it amounts to the property that every combination of relations must be in one of the eight sections. Secretly, you already know, for example, that the set of all real numbers satisfies closure under addition in the sense that no matter which two numbers you choose, by adding them together you obtain another number. The Warlpiri would not write the sections and operations in the same symbolic way that we have; however, from a practical point of view we can ask if the Warlpiri would have some large chart or record of data to use for determining all possible combinations. The answer is that this is not necessary, because all they would need to know is what happens to the original eight sections. In Example 4.4 above we could think of this as determining the section of $f^3m^{15}fm^2f$, starting from the relation $m = m(I)$, which is section 3. Thus, what we need is to know what happens to each of the eight relations when they are applied to each of the other seven relations, including what happens when we apply a relation to itself. In the next examples are more calculations that will lead to some conclusions.

Example 4.7 Calculations of the eight operations.

(a) Let us calculate the father of m^3, that is, the operation from section 7 (which is f) applied to the operation of section 4 (which is m^3): The mother of a person in section 4 is in section 1, and she marries a man from section 5, so the father of someone in section 4 is in section 5 (which is mf). We describe this as $f(m^3) = mf$.

(b) If we calculate m of m^2, we get m^3, and if we apply m again, we get $m^4 = I$. We can express this in another way: By calculating m^2 of m^2, we get I, so $m^2m^2 = I$.

Example 4.8 What does "I" represent as an operation?

We had decided, albeit arbitrarily, that I would represent section 1. As a mathematical operation, we have to think about what I means in a different way. We move from section to section by either applying m or f, or by marriage between a person in one section and a person in another. This means that if I, as an operation on the sections, were to move us to another section, then I would represent some other relation that we had not previously included in the kinship system. Thus, we want I to keep us in the same section until we apply either m, f, or consider marriage. We can check to see if the property that I keeps us in the same section is consistent with how we have determined relations between the sections using combinations of m, f, and marriage. For example, $I(I) = I$ expresses the idea that if we start in section 1, we remain in section 1. Another example is if we consider the operation of m^5. By using the associative property, we can write this as $m^5 = m(m^4) = m(I) = m$, but also as $m^5 = (m^4)m = I(m) = m$, and this expresses the fact that if we apply m to I or I to m, we always get back m. In fact, if we have any combination of operations that returns us to I, then we can think of that operation as indicating that $I(A) = A$, where A represents any combination of operations. In Western mathematical terms, we say that, as an operation, I acts as an *identity* because it leaves all other objects unchanged. For addition of numbers, zero is the identity because for any number a, $a + 0 = a = 0 + a$. For multiplication, the number 1 is the identity.

Example 4.9 What happens when we apply an operation that leads to "I" ?

We saw in Example 4.7b that $m^2(m^2) = I$. Here is another example of how that can happen. If we start with m^3 and apply m to it, we get $m^4 = I$, so $m(m^3) = I$. Likewise, $m^3(m) = I$. Comparing this with operations on numbers, we say that m and m^3 are *inverses* of each other because applying one to the other returns us to the identity I. For addition of numbers, the inverse of a given number a is its negative because $a + -a = 0$, and 0 is the additive identity. A specific example would be that –11.7 is the inverse of 11.7 because $11.7 + -11.7 = 0$. Here is another example from the Warlpiri kinship.

Example 4.10 Is there an inverse for mf?

If we apply m^3 to mf, we obtain $m^4f = f$, so m^3 is not an inverse of mf. On the other hand, if we apply mf to mf we get I, as you can check. Thus, the inverse of mf is mf.

Another example is that which we saw with fathers, the operation f; whenever we consider the father of a father, we return to whatever section we started in, so that means $f(f) = I$. In other words, the inverse of f is again f. As with the associativity and closure we can check *every* operation that represents one of the eight sections and it turns out that each one has an inverse.

We can see that if we have basic information about the relations that represent the eight kinship sections, then we can obtain a lot of additional information using properties such as associativity and inverses. In fact, the Warlpiri kinship system has a particular mathematical structure that we can identify in Western terms. That is the topic of the next section.

WARLPIRI KINSHIP AS A GROUP

I would like to summarize the properties of the Warlpiri kinship system. For reference, we can make a complete table of the Warlpiri kinship relations, as appears in Table 4.2. An important point in reading Table 4.2 is that *the order matters*. To calculate operations between relations using the table, *the relation in the row operates on the relation in the column*. For example, to determine mf of m^3, look for the row marked **mf**. Then read across the column headings until you come to the column marked $\mathbf{m}^3$. Where the mf row and the m^3 column intersect is $mf(m^3)$, which you should see is m^3f. Notice that if you read in the opposite order, that is, if you look at where the m^3 *row* intersects with the mf *column*, you will see f. This represents the operation of m^3mf, which is not the same as mfm^3. In the exercises I have asked you to verify some of the other entries of the table and look for inverses.

TABLE 4.2 Warlpiri Kinship

	I	**m**	**m²**	**m³**	**f**	**mf**	**m²f**	**m³f**
I	I	m	m^2	m^3	f	mf	m^2f	m^3f
m	m	m^2	m^3	I	mf	m^2f	m^3f	f
m^2	m^2	m^3	I	m	m^2f	m^3f	f	mf
m^3	m^3	I	m	m^2	m^3f	f	mf	m^2f
f	f	m^3f	m^2f	mf	I	m^3	m^2	m
mf	mf	f	m^3f	m^2f	mf	I	m^3	m^2
m^2f	m^2f	mf	f	m^3f	m^2f	mf	I	m^3
m^3f	m^3f	m^2f	mf	f	m^3f	m^2f	mf	I

Note: Apply the relation of the leftmost column to the relation in the top row to get the relation in the intersection of the row and column.

The summary of the Warlpiri kinship system, with Western mathematical terms in parentheses, goes like this:

- Every combination of relations always amounts to one of the eight original relations (*closure*).
- The associative property holds (*associativity*).
- There is a relation I that leaves all other relations unchanged (*identity*).
- For every relation there is some relation that combines with it to equal the relation I (*inverse*).

Now for the surprising part. In modern Western mathematical terms, any collection of objects for which there are operations that satisfy the four properties above (closure, associativity, identity, and inverse), is called a *group*. Group theory is a very abstract branch of Western mathematics. Nevertheless, the Warlpiri kinship system represents a group. In Ascher (1991: chapter 3), you can find an excellent description of the Warlpiri kinship system and more details regarding its structure as a group. For example, the Warlpiri kinship system is mathematically equivalent to the group formed by the eight symmetries (including the identity) one can form with a square (e.g., rotation by 90°).

CULTURAL ASPECTS OF WARLPIRI KINSHIP

A main theme of this book is to recognize that people from distinct cultures can understand the same mathematical concept even though they would interpret and describe that concept in entirely different ways. If we were to ask people from the Warlpiri culture to describe their kinship system, they probably would not use mathematical terminology to describe it. Objects that we think of as operations, properties, and so on would be described by the Warlpiri as *family relationships*. The importance they would put on their kinship system would have to do with the social and cultural obligations and consequences of the section to which a person belongs, not the mathematical properties of the eight kinship sections. Yet the Warlpiri understand many of the concepts we have examined from a mathematical point of view. For example, they would describe the closure property in terms of each person belonging to one and only one kinship section, and the associative property would be described in terms of how the sections of past, present, or future generations are determined.

OTHER RELATIONS

In this chapter we have discussed the cultural mathematics of a system of kinship relations. This is but one example of the many types of relations that can exist between people.

In Ascher (1991: chapter 3) and Ascher (2002: chapter 5) you can find descriptions of several other examples of relations between people, such as social relations of the Basque culture of Europe and the Borana culture of Africa.

Other systems of relations in cultures can occur within the context of rituals. In such cases some persons often have special roles and duties, and their relations with others involved in the ritual process can be complex. Yet another context of systems of relations can occur within the context of a culture's perceptions of nature, deities, and related themes. Although these are not relations between people, their structure can be complicated and mathematical. We saw earlier how the Ojibwe of North America use the number four in a ritualistic way. If you read Closs (1986: chapters 6 and 7) you will find that there are numerous descriptions of relations between humans and nature that are expressed by the Ojibwe in terms of powers and combinations of the number four. This ritual use of the number four represents a mathematical aspect of relations between people and the physical or spiritual world.

EXERCISES

Short Answer

4.1 Why is the Warlpiri kinship system described in terms of mothers and fathers instead of daughters and sons? Would the system be mathematically the same if the relations were described in terms of daughters and sons?

4.2 Explain the following relations from the Warlpiri, using Figure 4.1, as necessary:

(a) A boy is in section 6. In which section is his father?

(b) In which section is the boy's sister?

(c) In which section is the boy's cousin? If there is more than one possibility, explain each one.

Calculations

4.3 A girl is in section 3. Explain the following relations, using Figure 4.1, as necessary:

(a) Determine the possible sections in which a great-grandmother of the girl could be.

(b) Determine the possible sections in which a great-great-grandfather of the girl could be.

4.4 A boy is in section 5. Determine the possible sections in which a great-great-grandmother of the boy could be.

4.5 Determine the other three patricycles of the Warlpiri kinship system; that is, starting with section 1, obtain the patricycle corresponding to this section, then repeat for sections 2 and 3.

4.6 **(a)** Suppose we define two functions f and g to be equal on an interval $[a, b]$ to mean that for each x in $[a, b]$, $f(x) = g(x)$. Show that this defines an equivalence relation by showing how the reflexive, symmetric, and transitive properties are always true.

(b) Now suppose we define two functions f and g to be "equal" on an interval $[a, b]$ to mean that $f(x) = g(x)$ except for a finite number of values of x in $[a, b]$. Show that this defines an equivalence relation by showing how the reflexive, symmetric, and transitive properties are always true. You may have to figure out some properties of finite sets if you do not already know them.

4.7 Define two integers a and b to be equivalent if they are both divisible by three. Explain how this is an equivalence relation on the set of all integers. Would the property still be true if the divisibility were by an integer other than three? Explain.

4.8 Calculate the following operations from the Warlpiri kinship system. Explain your steps. You may find that using the associative property helps in some cases.

(a) f^{14}. **(b)** fm^7f. **(c)** $(m^2f^2)(fmf)$ **(d)** $m^3fm^5f^3m^7f$.

4.9 Determine the inverses of the following Warlpiri operations. Explain your steps.

(a) m^3f. **(b)** mf. **(c)** f^3mf. **(d)** $m^6f^4m^3f^3$.

4.10 Suppose instead of using section 1 as I we were to use section 6 as I.

(a) Determine the entries of Table 4.1 that we would obtain with this new choice of I.

(b) Determine the entries of Table 4.2 that one would obtain with this new choice of I. The identity and associative properties will help reduce the number of actual calculations you must make.

Exercises 4.11–4.16 assume some knowledge of group theory.

4.11* Refer to Exercise 4.10 above. Prove that the Warlpiri kinship system is a group with the choice of $I = 6$, and that we obtain a group regardless of the choice of I.

4.12* **(a)** Prove that for any nonzero integer k, $f^{2k}m = m$. Prove that, in general, if A is any relation representing one of the eight Warlpiri sections, then $f^{2k}A = A$.

(b) Determine the form of a nonzero integer N for which $m^N A = A$ holds, where A is as in (a).

4.13* Is the Warlpiri group cyclic? If so, determine the generators of this group.

4.14** **(a)** Give a *cultural* explanation for why the Warlpiri group is nonabelian. (Recall from abstract algebra that a group with an operation $*$ is *abelian* if for all elements a and b, one has $b * a = a * b$; the operation is *nonabelian* if the equation $b * a = a * b$ does not hold for all elements a and b.)

(b) Prove that it is impossible to construct a nonabelian group having only three elements. In general, it can be shown that every group with five or fewer elements must be abelian.

(c) Based on your results from parts (a) and (b), explain whether or not a Warlpiri-like kinship system having five sections would make sense.

4.15* Determine all proper subgroups of the Warlpiri kinship group.

4.16* Prove that the Warlpiri kinship system is isomorphic (in the group theory sense) to D_4, the dihedral group of symmetries of a square.

Essay and Discussion

4.17 Even though the Warlpiri system is closed under the operations described in this chapter, there are cases in which it would not be clear in which section a person should belong. Write or discuss several such possibilities and give explanations for how the person's section could be decided by the Warlpiri. Keep in mind the cultural aspect of such decisions, too.

4.18 Investigate a kinship system of a culture other than the Warlpiri (for Western culture, see below). Construct tables of relationships similar to those for the Warlpiri. If the kinship system does not follow a mathematical pattern similar to what we have discussed in this chapter, discuss other aspects of the system and how they relate to mathematics. For example, you can describe the answer to the question of how a person determines relationships with others in the culture.

4.19 Write/discuss Western kinship. Explain terms such as "first cousin removed," "second cousin," "half sister," "half brother," "stepmother," and "stepfather." Explain examples of each of these terms, either from your own experience or by constructing a fictitious family. Finally, explain the mathematical aspects of this kinship relation system. Note that even though it does not represent a group like the Warlpiri, there are mathematical considerations; see Chapter 1 on Bishop's six.

Additional Comments

- Most of my writing of this chapter has been inspired by chapter 3 of Marcia Ascher's *Ethnomathematics* (1991). She lists many references on the mathematics of kinship, more about the Warlpiri culture, and so forth, for those who wish to investigate this topic further.
- In the interest of keeping the length of the discussion shorter, I have focused only on kinship relations. The general idea of relations between people and objects, and the mathematics that can be part of such relations, is a big topic. In the last section on "Other Relations" I tried to give some idea of the possibilities along with references for the interested.

5

ART AND DECORATION

Motivational Questions: Before you begin reading this chapter, take some time to think about your responses to the following questions.

- *Is there anything mathematical about creating art? Describe examples you can think of.*
- *Find some work of art that has a repeating pattern in it. How was the pattern made?*
- *What does it take to become highly skilled as an artisan, particularly in textile art?*
- *What does it mean for a figure to be vertically symmetric or horizontally symmetric?*
- *Do you know someone who creates textile art, such as quilts, weaving, embroidery, or knitting? If so, do you think that person is using mathematics to make the art? Explain.*
- *What is the mathematical explanation for the fact that if you turn the letter* E *upside down you get the same letter but this is not the case for the letter* F*?*

Introduction to Cultural Mathematics: With Case Studies in the Otomies and Incas, First Edition.
Thomas E. Gilsdorf.

INTRODUCTION

Art and decoration is all around us. Some of it is manufactured and some of it might be handmade. We will see in this chapter that creating artwork, especially textile art, typically utilizes mathematical concepts and thinking. This topic provides an excellent example of an activity that we may first think of as a hobby, yet turns out to be quite mathematical. The focus in this chapter will mainly be on textile art made by traditional artists; however, many of the same thoughts can be applied to other art forms, and even to art that is created by machines. By *traditional artist* I am referring to artists that create their work using techniques and knowledge that have been part of that person's culture for quite some time, usually generations. Before we begin, some general observations about creating traditional art ("traditional" here means that it has been preserved for generations in the culture) will motivate our mathematical considerations.

First, to become a traditional artist is typically a life-long process. A common situation is one in which there are several generations of people involved in creating projects. In Ascher (1991: chapter 6, section 3), you can find some details of what is necessary to become a master craftsperson in the Maori culture of Australia. Later in this book when we discuss the Otomi and Inca cultures, you will also see that the process of becoming a traditional artist is a long one.

Second, the work done by traditional artists often has important cultural meaning. In many cultures, for example, the color or pattern of the outer clothing (which usually consists of textiles created by artists) worn by a person indicates such things as the place where that person lives and the social status of the person, and identifies the person as having a special role in the community. Moreover, the artists who create these works are typically treated with honor and respect in the community. See Ventura (2003: chapter 3) for a detailed explanation of these ideas within the context of the Mayan culture, as one example.

Third, traditional artists often create the patterns of their work completely by memory. If you have ever had a chance to visit a market in which artwork is being sold, you may have noticed that aside from selling products, the people are creating more artwork to be sold. If you look closely, you will notice that they are not looking at diagrams or pictures as they create that art. Also, as they are creating, they are not using rulers to measure. In short, the artist must keep track of a lot of information from memory. In the case of textile artists, part of that information includes keeping track of many counts of threads, as well as symmetric properties.

The previous considerations already give us the feeling that creating traditional art is not so simple, and in fact we will see that mathematics is being used throughout most of the process. To get ready for the discussion on textile art, we will first think about some mathematical notions of symmetry from a Western context.

SYMMETRIC PATTERNS

Take a look at this: ♈. If we were to draw an imaginary line down the middle of this figure, then the part on the right side and on the left side would look the same, though facing in opposite directions. Another way to think of this is that we could draw an imaginary line down the middle of the figure and then reflect the figure from the right to the left (or left to right), and we would see that we get exactly the same figure back. This is called *vertical symmetry*. The Western definition would be something like this: A figure has vertical symmetry if whenever you reflect it across a vertical line in the middle of the figure, you get the same figure back. Now let us try this figure: ❧. If we draw a line down the middle of this figure, then the left and right hand sides do not look alike. Equivalently, if we reflect the figure across a vertical line that passes through the middle of the figure, we get this: ☙. This is not the same figure as the original, and we conclude that this figure is not vertically symmetric.

Here are other kinds of symmetry: A figure is *horizontally symmetric* if when we reflect it across a horizontal line drawn through its center, we obtain the same figure back. This figure is horizontally symmetric: ▶. This one is not: ▼, because if you reflect it across a horizontal line through the middle of it, you get ▲. Next, there is *rotational symmetry*, in which we get back the same figure by rotating the figure about a point in the center of the figure. Of course, we have to clarify how much the figure is rotated. The most common choices that we will consider would be to rotate either 90° or 180°, though any other angle between 0° and 360° is possible. I will always assume the rotation to be *counterclockwise*. Thus, we can talk about 90° rotational symmetry or 180° rotational symmetry. Consider this figure: ∅. If you imagine putting a tack in the center of the figure and rotating it 180°, you will get back the same figure. So, it is rotationally symmetric through 180°. You can check that ∂ does not satisfy this rotational symmetry. Finally, we can consider figures that satisfy more than one type of symmetry, or that satisfy one kind but not another. For example, ▣ satisfies all of the symmetry properties we have discussed. That is, if you rotate it 90°, or 180°, or reflect it across a vertical or horizontal line, you always get back the same figure. On the other hand, $\int$ satisfies 180° rotational symmetry, but none of the others. You can check this by rotating it 90°, reflecting it across a vertical line, or across a horizontal line. You will see that you do not get the original figure back. Now, it is time for you to try creating some symmetric objects.

Example 5.1 Complete the following pattern so that it has horizontal symmetry: ▼. Is there more than one way to complete this pattern?

Notice that one way to complete the pattern and have horizontal symmetry is this: ◆. Another way is this: ⧗.

Example 5.2 Complete the following pattern so that it has vertical, horizontal, and 180° rotational symmetry: ☺. Is there more than one way to complete this pattern?

Notice that the figure already has vertical symmetry. One way to complete the pattern and have all three symmetry properties is this:

Another way would be to copy the figure three more times so that it has a cloverleaf pattern to it, with each face oriented such that the smile points to the center of the figure.

Now for an important question: *How did you create the symmetry in the previous two examples?* As you think about it, you will realize that, for instance in Example 5.1, what you must do is to make a copy of the original figure that is *oriented in the opposite vertical direction*. That is, the copy must be "upside down." Similar observations can be made regarding Example 5.2. You can do these things because you understand what it means to have horizontal, vertical, and 180° rotational symmetry. This is how traditional artists typically create their art, too. We will see in Chapter 9 that traditional textile artists of the Otomi culture create half of the pattern of their projects, then orient the figures properly to make the other half so that the final image has the desired symmetry properties. Those artists understand the symmetry concepts they are working with! Also, it is worth reminding ourselves that the artist typically creates these figures and patterns from memory.

Although there are other types of symmetry that we could consider (such as reflecting across lines that are neither horizontal nor vertical), the above examples suffice for our discussion here. In the exercises you can find questions about the different combinations of symmetry. In a later section we will discuss strip patterns in which we consider symmetry in a more abstract setting, and we will see other properties there. Meanwhile, we return to the discussion of art. Our goal here is to investigate some examples of art to determine which kinds of symmetry properties they have, and how such figures were created by the artist to make the work. In what follows are several examples of art work from distinct cultures.

Example 5.3 Dakota beadwork, shown in Figure 5.1.

Notice that if we reflect the figure across a vertical or horizontal line, we get the same figure back. Likewise, if we rotate the figure 180°, we also get the same figure back. Thus, this figure satisfies vertical, horizontal, and 180° rotational symmetry. By the way, in the exercises you will see that if a figure satisfies two of the three properties of vertical, horizontal, and 180° rotational

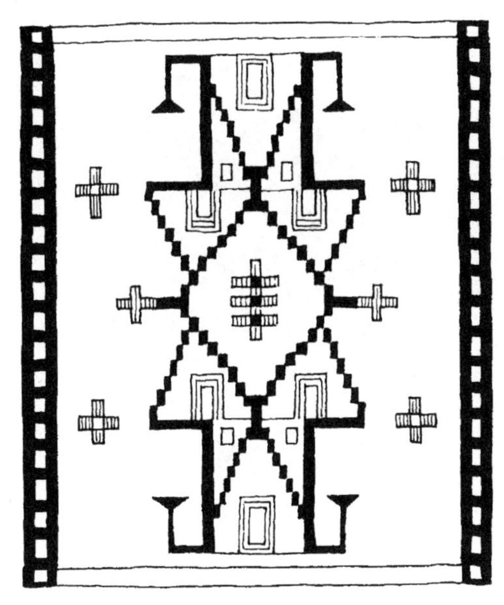

Figure 5.1 Beadwork from the Dakota culture. (Courtesy of *Appleton's American Indian Designs*, Le Roy Appleton, Dover Publications, 2003, image 084.)

symmetry, then it must satisfy the third property. In this case, once we knew that the figure satisfies horizontal and vertical symmetry, we can conclude that it also satisfies 180° rotational symmetry as well.

Example 5.4 Chinese art, shown in Figure 5.2. You can check that it only satisfies vertical symmetry, and none of the other symmetries.

Example 5.5 African art, shown in Figure 5.3. You can check that it only satisfies horizontal symmetry, and none of the other symmetries.

Example 5.6 For Figure 5.4, determine the smallest positive angle A, such that the figure has $A°$-rotational symmetry; that is, if we rotate the figure by A degrees, we obtain the same figure back. Explain what the symmetry property implies about the mathematical knowledge of the artist who made the figure.

One way to determine the angle is to draw some line segments from the center of the figure through the parts of the figure that are near the edge of the figure, then measure the angle between the segments. Note that, in order to have the angle measurement be accurate, we would have to draw the segments so that each one crosses the same part of the figure. For example, each segment could go from the center through the tip of the tail of each of the fish figures. Another

Figure 5.2 Chinese artwork. (Courtesy of *Treasury of Chinese Designs*, Stanley Appelbaum (ed.), Dover Publications, 2006, image 204.)

Figure 5.3 African artwork. (Courtesy of *African Tribal Designs*, Geoffrey Williams (ed.), Dover Publications, 2004, image 165.)

way would be to look carefully at the figure and notice that there are four aspects to the figure, each of which was created to be of equal distance from the adjacent figures. In other words, in this figure, the four fish, the small circles, and the center figure have been created so that each one is the same distance (and is the same size as) from the next figure. Because a circle consists of 360°, we conclude that the angle A that we seek is one fourth of 360°, or $A = 90°$. You can check this by rotating the figure 90° to see that you get the same figure back.

Now, to explain what this implies about the artist. We can say that the artist understood that concept of rotating a figure by one fourth of a circle in order to obtain the required symmetry. Observe that the artist probably specifically

Figure 5.4 Turkish artwork. (Courtesy of *Traditional Turkish Designs*, Azade Akar, Dover Publications, 2004, image 303.)

concentrated on the rotational aspect of the symmetry because the figure satisfies neither vertical nor horizontal symmetry properties. Finally, it is worth noting that the artist would not necessarily describe the rotation in terms of degrees as we have done here. Let us look at another example.

Example 5.7 For Figure 5.5, determine the smallest positive angle A such that the figure has $A°$-rotational symmetry; that is, if we rotate the figure by A degrees, we obtain the same figure back. Explain what the symmetry property implies about the mathematical knowledge of the artist who made the figure.

As with the previous example, we could measure the angle directly. We could also note that the there are three equally spaced flower figures of the same size in the larger figure. Thus, the angle between each of the smaller flower figure must be one third of 360°; hence, $A = 120°$.

To describe what this implies about the the artist's mathematical knowledge, we can observe that the artist understood the concept of creating each flower figure in the proper orientation so that rotating the figure one third of a revolution would give back the original figure. Also, note that the artist would not necessarily have used degrees to describe this rotation.

With these examples, it is time for us to move on to the next topic of artwork, strip patterns.

Figure 5.5 Japanese artwork. (Courtesy of *Japanese Motifs and Designs*, Joseph D'Addetta, Dover Publications, 2007, image 134.)

STRIP PATTERNS

A common expression of art is what we will call (see Ascher, 1991, or Zaslavsky, 1999) *strip patterns*. A strip pattern is one in which a specific figure is repeated several times. It can be vertical or horizontal. On a ceramic piece of art that is either spherical or cylindrical, this would mean the figure is repeated around the side of the figure. In the case of a textile that is rectangular in shape, the figure would be repeated across the length or width of the textile. However, a strip pattern does not have to repeat all the way around or across a piece of art; it may only appear on part of the art. Look around your world. You will see strip patterns in many places, such as around the top of a room, in a tablecloth, on the glass from which you drink, on your clothes, or on your blankets.

In Western mathematics, properties of symmetry have been studied extensively by Russian crystallographers (see Ascher, 1991: 181–182). Although people from many cultures have observed properties of symmetry, it will be useful to us to categorize patterns on a strip in Western terms. Our beginning assumption about such patterns is that they repeat indefinitely. In this sense, we are assuming that the pattern satisfies the property that if you translate it to the left or to the right, you get back the same pattern. It turns out that the pattern of only translation can be considered as a group. Recall that a *group* is a set of objects with a defined relation (an "operation") that satisfies certain properties such as having an identity, the associative property, and that each object has an inverse, as we saw in Chapter 4 in the Warlpiri kinship system. More mathematical details of symmetry patterns as groups (so-called symmetry groups) can be found in Ascher (1991: chapter 6). Now, if we start with a pattern that only satisfies translation and change it so that it has horizontal symmetry, then we have introduced a second symmetry property that is not satisfied by the translation-only pattern. This implies that the group formed by the pattern that has horizontal symmetry must be different from the group formed by the translation-only pattern. We can experiment with other symmetries such as vertical symmetry, and each time we obtain a pattern on a strip that cannot occur in the previous pattern, we will get another group. It may seem that there will be many groups, but it turns out that there are only seven possible groups of patterns on strips that can be created in this way. I will use the vocabulary of Zaslavsky (1999) by referring to these seven possibilities as *types*, that is, type 1, type 2, and so on, up to type 7. Table 5.1 shows the seven types and examples of each.

As we look at various examples of artwork, we will see these patterns appearing as geometric forms instead of letter-produced. Here are some more details concerning the seven types. Type 1 is the pattern that only repeats as you translate, that is, as you move from left to right or right to left. There is no other symmetry property present. Types 3, 4, and 6 are the vertical, 180° rotational, and horizontal symmetry properties that we have already seen. Type 5 represents both 180° rotational and vertical symmetry, and type 7 represents both vertical and horizontal symmetry. Hence, for example, if you take the pattern in type 5 and rotate it 180° about its center, or if you reflect it across a vertical line in its middle, you will get the same pattern back. The new symmetry that appears is the *glide reflect* symmetry. This symmetry is defined as follows: Take the figure of type 2 and translate it to the right, then reflect it horizontally. You should get back the same figure. Finally, it is worth mentioning that each of the types includes translation as one of its properties. Now let us look at some examples of strip patterns.

Example 5.8 A glide reflect example, shown in Figure 5.6.

To see the glide reflect symmetry, look at the flower image that is second from the left (it is oriented upward). Now imagine that you slide that part of the figure to the right, the length of one flower figure, then reflect it horizontally (i.e., flip it upside down). That second flower figure would now be exactly where the third flower figure is. Notice that this symmetry applies to each of the flower figures; that is, if you "glide" one such figure to the right one length (of the flower), and reflect it horizontally, you get exactly the next flower pattern. Observe that as a strip pattern, the translation property (always) holds in the sense that as you translate (slide) the pattern to the right (or left), it repeats. Finally, as a figure that satisfies only the glide reflect and

TABLE 5.1 The Seven Types of Symmetry Properties on a Strip

Symmetry Group	Example	Name
Type 1	pppp . . .	Translation (only)
Type 2	$^{q}d^{q}d$	Glide reflect symmetry
Type 3	YYY . . .	Vertical symmetry
Type 4	$^{b}q^{b}q$	180° rotational symmetry
Type 5	$bd^{pq}bd^{pq}bd$	180° rotational and vertical symmetry
Type 6	dddd . . . qqqq	Horizontal symmetry
Type 7	bbdd . . . pqpq	Vertical and horizontal symmetry

Figure 5.6 American folk art: embroidery from New York, 1807. (Courtesy of *American Folk Art Designs*, Joseph D'Addetta, Dover Publications, 2006, image 091.)

translation properties, this strip pattern is a type 2 pattern, according to Table 5.1.

Example 5.9 Artwork from the Aguada culture of what is now Argentina, shown in Figure 5.7.

The first thing to note is that this figure is another strip pattern in the sense that the patterns repeat as you follow it from right to left (or left to right). The pattern satisfies the translation property, as do all figures that we are calling strip patterns. As you look at the figure, you can see that for an imaginary vertical line through the middle of the figure, the images reflect across it, so it has vertical symmetry. On the other hand, reflecting horizontally would put the eye patterns on the bottom so it does not have horizontal symmetry. You can check that the symmetry patterns in this figure consist of translation and vertical symmetry. Thus, it is in symmetry group type 3. In Figure 5.8 we can see examples of the seven strip pattern symmetry types.

Let us look at one of the patterns and explain which type of strip pattern symmetry it satisfies. Consider the strip pattern in Figure 5.8d. If we look closely at it, we can see that it if we rotate this page 180° and look at the pattern in Figure 5.8d, it looks the same. So, the pattern satisfies 180° rotational symmetry, and is type 4. You can look at the other six examples and deduce the

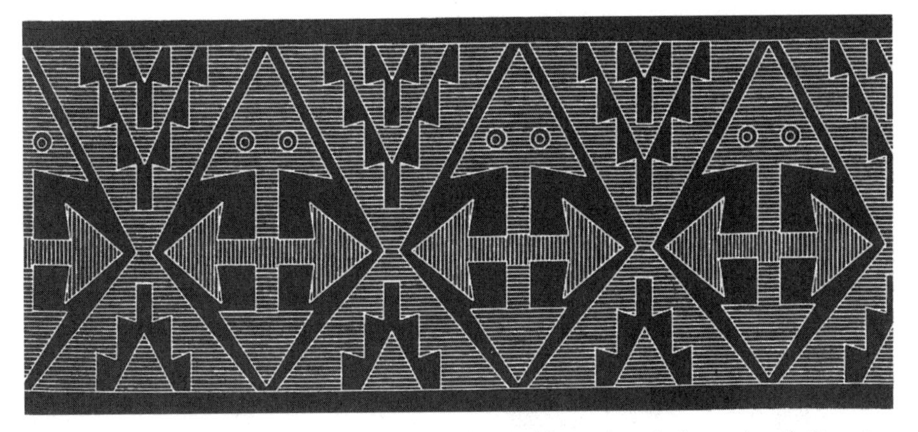

Figure 5.7 Artwork from the Aguada culture. (Courtesy of *Argentine Indian Art*, Alejandro Eduardo Fiadone, Dover Publications, 1997, p. 16.)

type of symmetry pattern each is based on, as described by types 1 through 7. Notice the interesting variety of artwork represented, of cultures from several parts of the world.

Figure 5.9 shows the relation between all seven types of symmetry on a strip. This will give you a feel for how the different types relate to each other mathematically.

There is much more to patterns on a strip if we include *color* as a factor. Indeed, you have probably seen examples of patterns in your world where colors alternate or appear in some predictable way.

The main idea is how a color appears with respect to whatever symmetry properties that are present. One possibility is that the colors always appear in the same part of the pattern. This is the case if the bottom right corner of the figure is blue every time the figure repeats, for instance. Another possibility is that the color alternates with the symmetry; that is, each time the figure repeats, colors are alternated. So, including color makes the situation more complicated or, as I would prefer to see it, more interesting. In fact, if we include color as a consideration, the number of symmetry groups expands from seven to 24. There are more details in Ascher (1991: chapter 6). I will discuss a few such possibilities here.

Example 5.10 Motif from a Turkoman carpet, shown in Figure 5.10.

Before considering color, we can determine the symmetry properties of the figure. Indeed, if we reflect it across a vertical or horizontal line, or rotate it 180°, we get the same figure back, so it satisfies all of these symmetry properties. Now, if we include the aspect of color, notice how the colors alternate between shaded and nonshaded parts that correspond to each other. For example, for the nonshaded triangle in the upper left part of the figure, if we

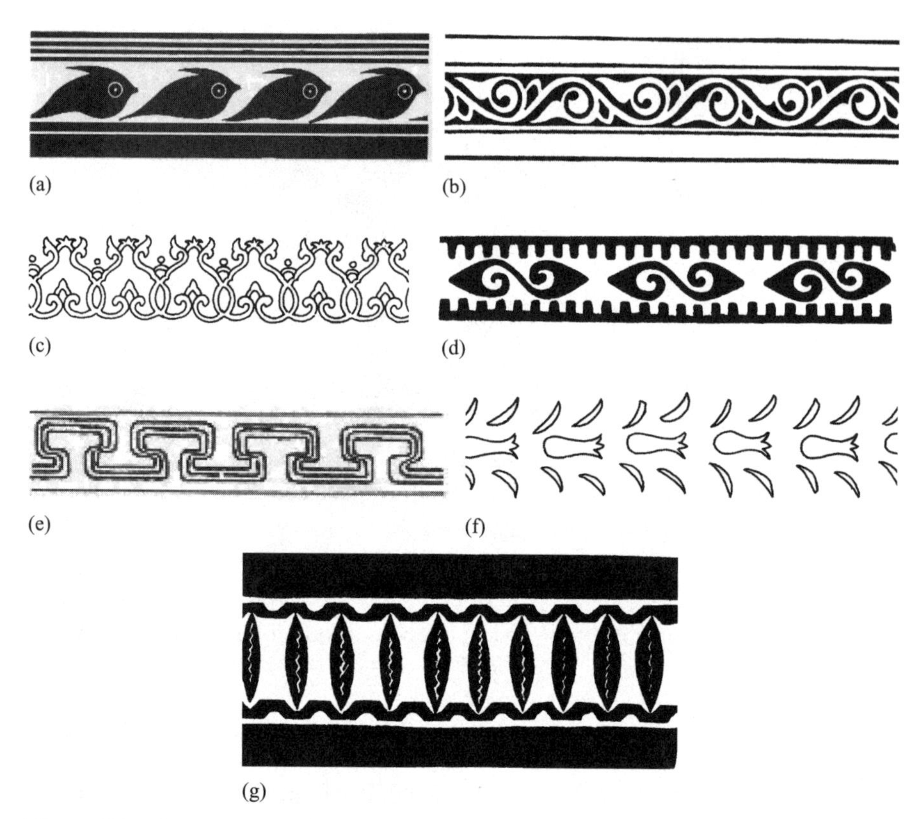

Figure 5.8 Cultural examples of the seven symmetry patterns on a strip. (a) Type 1: a prehistoric ceramic design from ancient Ecuador; (b) type 2: a Persian stucco border; (c) type 3: a Chinese design; (d) type 4: a prehistoric design from ancient Mexico; (e) type 5: a prehistoric design from the Middle Mississippi Valley cultures; (f) type 6: American folk art design from Connecticut, 1790; and (g) type 7: African design (culture not specified). (Images courtesy of Dover Publications: (a) Shaffer (1979: plate 164); (b) Dowlatshahi (2007: image 126); (c) Appelbaum (2006: image 116); (d) Enciso (2004: plate 141); (e) Naylor (1975: 28); (f) D'Addetta (2006: plate 140); and (g) Williams (2004: plate 090).)

apply the vertical symmetry to the figure, the triangle then corresponds to a triangle (oriented properly to preserve the vertical symmetry) in the upper right part of the figure, and that triangle is shaded. Also, the background in the upper right part of the figure is shaded, but when vertical or horizontal symmetry is applied, the corresponding background for that part of the figure is not shaded. The alternating colors emphasize the symmetry properties in this case.

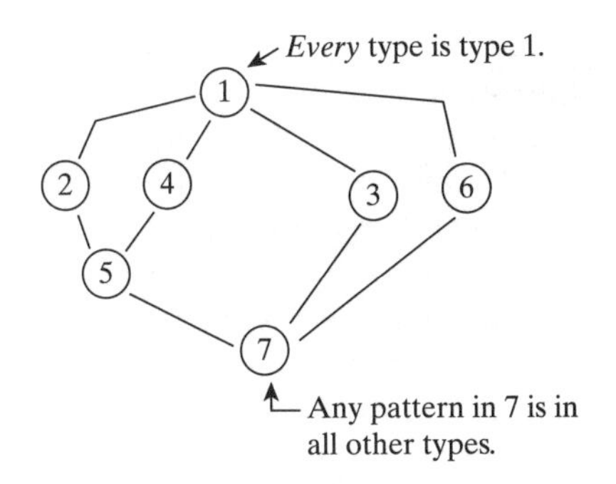

Figure 5.9 The seven symmetry types for strip patterns.

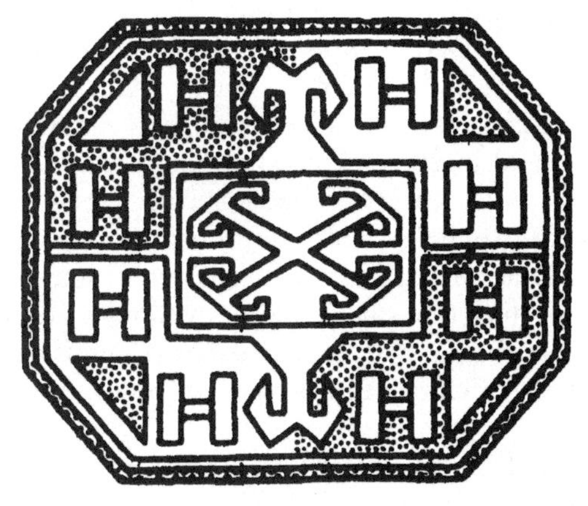

Figure 5.10 Artwork from a Turkoman carpet (from ancient Persia). (Courtesy of *Persian Designs and Motifs*, Ali Dowlatshahi, Dover Publications, 2007, image 078.)

CULTURAL ASPECTS OF ART

In this section we will focus on the cultural context of artwork. Symmetry is a good starting place. What is its cultural significance? A more general consideration of symmetry is that it represents a human attraction to figures that represent some kind of organization of space. See Ascher (1991: chapter 6) for some general interpretations of what art says about humanity. At any rate, for whatever reason, humans find certain kinds of patterns attractive, including ones with symmetric patterns. This is a good beginning. The next step is to look at what art, and the mathematics of it, can tell us about a specific culture. In particular, if certain mathematical properties consistently appear in the artwork of a culture, then this implies certain cultural tendencies. This is a remarkable example of mathematics giving us an insight into culture. In what follows, we will delve into some details.

Now that you have seen several examples of artwork, recall that we observed after Examples 5.1 and 5.2 that *the artists who make such art understand the symmetry concepts*. To answer the question of the cultural significance of artwork for some specific cultures, we can think about how the work was made. Starting with the Dakota beadwork of Figure 5.1, notice that in order for the figure to have horizontal symmetry, the design in the bottom half of the figure must match that of the upper half. To accomplish this, the design of the bottom must be created in the *opposite orientation* of the upper half. Thus, the artist must visualize that change in orientation in order to have the symmetry property be satisfied. This means he or she applies an *understanding* of the particular symmetry concept. The same reasoning can be applied to all of the symmetry properties that we have discussed.

Notice something else that is important: The artist probably *would not describe the symmetry properties in mathematical terms*, at least not in a way that we might be accustomed to. Think back to Chapter 1, and recall that people from distinct cultures can understand similar mathematical concepts but would describe and interpret them very differently. For example, recall that in the cultural context of artwork, the specific designs often have social or religious significance. In his or her description of the design and its patterns, the artist would include those factors of social importance, not mathematical terms. A particular pattern that satisfies some symmetry property might reflect a cultural viewpoint of the environment, ritual or religious beliefs, the universe, or many other factors. The description that follows shows this kind of connection.

Consider traditional Celtic artwork, such as shown in Figure 5.11. Take a look at the figures and observe what kind of symmetry or other properties they have or do not have.

Now that you have looked at the figures, consider the following quote regarding Celtic art in Green (1996: 209): "Rather different manifestations of the same basic phenomenon include triplication of horns on bull-figurines, . . . , and the three horns on a boar-figurine are further examples of the ubiquity of triplication in various forms." She goes on to state a few lines later, "In the vast majority of instances where triplism is present, it is evident that the number '3' is important and so is likely to have possessed powerful symbolism for the Celtic people." Thus, the number three carries a cultural importance in traditional Celtic culture. Do you see the connection with the figures of artwork? The figures both show patterns that repeat in groups of three. Although this is not a rigorous study of the frequency in which the number three appears in Celtic art, we can still conclude that in many instances, an importance of the number three is expressed in the art. This represents an example of how our mathematical observations of art can give us insights into the cultural aspects of art.

Figure 5.11 Examples of Celtic art. (Courtesy of *Celtic and Norse Designs*, Amy Lusebrink and Courtney Davis, Dover Publications, 2006, images 216 and 210.)

OTHER FORMS OF ART

Throughout most of this chapter the main types of artwork under consideration have been of the form of textiles or patterns on textiles. There are, of course, many other expressions of artwork, and the mathematical information contained in them could also be investigated. For a glance into three-dimensional considerations, consider Figure 5.12, an ancient ceramic vessel from what is now Ecuador. Other types of three-dimensional artwork would include baskets, and you can find many examples of such artwork in Gerdes (2008), for instance.

The artifact shown in the figure has a symbolic representation of grain, and as you can see, there is vertical symmetry present in the form of a strip pattern that goes around the vessel. Notice how the strip pattern becomes three-dimensional, and also notice how the artist who made the vessel must have known how to create a fixed number of repeating figures from the strip pattern so that once created on the vessel, there are no partial patterns. Moreover, the artist could create one kind of strip pattern around the bottom of an object, a different kind of strip pattern around the top, and yet a third type of symmetry pattern (rotational symmetry, as we have seen in Example 5.5) on the bottom.

We have seen several aspects of art and mathematics, including cultural contexts. Now it is time for you to think more about this topic via the exercises.

182
Grain.
Black. 16 × 14 mm.

Figure 5.12 Artwork from ancient Ecuador. (Courtesy of *Indian Designs from Ancient Ecuador*, Frederick W. Shaffer, Dover Publications, 1979, plate 182.)

EXERCISES

Short Answer

5.1 **(a)** Give an explanation for why turning the letter E upside down gives us the same letter, while this is not the case for the letter F.

(b) If you turn the letter S upside down do you get the same letter back? What if you rotate it 180° (using the center of the letter)?

(c) Compare parts (a) and (b) in terms of symmetry properties discussed in this chapter. You may wish to repeat this exercise with other letters of the alphabet or numerical symbols.

5.2 **(a)** Rotational symmetry was described by imagining the rotation about the point that is in the center of the figure. If we were to rotate about a point that is not in the center of the figure, how would the description of rotational symmetry change, if at all? Examine some of the figures from this chapter (or from elsewhere) and describe rotational symmetry from this different perspective.

(b) What cultural observations can you make about a culture that creates figures in which the rotational symmetry has to do with rotation about a point that is not in the center of the figure? Do you know of any specific examples of a culture that does this?

5.3 Describe some patterns that you have seen inside a public building such as a school or government building. You may have to be quite observant, looking at such things as carpeting or woodwork. What kind of symmetry properties do the patterns exhibit? What cultural conclusions can you draw about the kind of patterns are used in such buildings? If there were only a few patterns you could find in the building, what does this imply?

Calculations

For Exercises 5.4–5.9, describe which of the symmetry properties (vertical, horizontal, 180° rotational) is satisfied by the figure (there may be more than one property). Explain how specific parts of the figure are oriented so that the symmetry property holds.

5.4 **(a)** Celtic/Norse design, shown in Figure 5.13.

(b) Considering that Figure 5.13 is from the Celtic/Norse culture, describe how the importance of the number three is expressed in the figure.

5.5 Painted designs and quill work on a skin coat, as shown in Figure 5.14. The image can be interpreted as four men hunting three animals that have antlers. Can you explain the cultural importance of this artwork?

5.6 Winnebago beadwork, as shown in Figure 5.15.

5.7 African artwork, as shown in Figure 5.16.

5.8 Iranian textile art, as shown in Figure 5.17.

5.9 Huichol textile art, as shown in Figure 5.18.

Figure 5.13 Celtic/Norse design. (Courtesy of *Celtic and Norse Designs*, Amy Lusebrink and Courtney Davis, Dover Publications, 2006, image 010.)

Figure 5.14 Skin coat from the Métis culture of central Canada and the northern United States. (Courtesy of *North American Indian Designs for Artists and Craftspeople*, Eva Wilson, Dover Publications, 1984, plate 53.)

Figure 5.15 Beadwork from the Winnebago culture (Midwest North America). (Courtesy of *Appleton's American Indian Designs*, Le Roy Appleton, Dover Publications, 2003, image 124.)

Exercises 5.10–5.16 require knowledge of abstract algebra and proofs.

5.10* *(Requires some ideas of mathematical proofs.)* Consider the three kinds of symmetry for strip patterns: vertical, horizontal, and 180° rotational. Prove that if a figure satisfies any two of them, then it satisfies all three, and use this information to explain why type 7 does not include 180° rotational symmetry in its description.

5.11* Which, if any, of the seven types of strip patterns are cyclic? Prove your assertions.

Figure 5.16 African artwork (culture not identified). (Courtesy of *African Tribal Designs*, Geoffrey Williams (ed.), Dover Publications, 2004, image 026.)

Figure 5.17 Textile art (carpet) from northern Iran, 1800s. (Courtesy of *Persian Designs and Motifs*, Ali Dowlatshahi, Dover Publications, 2007, image 090.)

5.12* Determine which of the seven types of strip patterns are subgroups of other types. Explain your conclusions. Create strip patterns, other than those in the chapter, that distinguish the groups from each other.

5.13* Determine if the group type 2 is abelian (see Exercise 4.14 for the definition of abelian). Determine if the group type 5 is abelian. Explain how

Figure 5.18 Textile art on a neckpiece from a woman's shirt, Huichol culture from present-day Mexico. (Courtesy of *Mexican Indian Folk Designs, 252 Motifs from Textiles*, Irmgard Weitlaner-Johnson, Dover Publications, 1993, plate 23.)

the figures from Table 5.1 are changed under the operations with which you are working.

Exercises 5.14–5.16 require knowledge of symmetry groups of geometric figures, defined using the isometries under the operations described as follows:

σ_0: identity; σ_1: horizontal reflection; σ_2: vertical reflection; σ_3: rotation, 180°; σ_4: rotation, 90°; σ_5: diagonal reflection (lower left to upper right); σ_6: diagonal reflection (upper left to lower right); σ_7: rotation, $\alpha°$ (specify α).

5.14* Let the set S be a rectangle with sides of length l and w, with $l \neq w$.

(a) Prove that neither σ_6 nor σ_7 are in the symmetry group of S by explaining why σ_6 and σ_7 do not preserve distances on S.

(b) Determine the symmetry group of a rectangle (not including squares). Explain your solutions in terms of symmetry properties and the operations σ_i described above.

(c) Prove that the symmetry group of part (a) is a subgroup of the dihedral group D_4 (isometries of a square).

5.15* **(a)** Determine the symmetry group of an isosceles triangle. Explain your solutions in terms of symmetry properties and the operations σ_i described above.

(b) Is the symmetry group of part (a) a subgroup of the dihedral group D_3 (isometries of the equilateral triangle)? If so, prove it. If not, explain why not.

5.16* **(a)** Determine the symmetry group of the figure in Exercise 5.8. Explain your solutions in terms of symmetry properties and the operations σ_i described above.

(b) Is the symmetry group of part (a) abelian? If so, prove it. If not, explain why not.

For the figures shown in Exercises 5.17–5.19, do the following:

(a) *Create a new figure from the original figure so that it has the desired symmetry.*

(b) *Explain how you oriented each piece of the figure in order to create the symmetry, and explain how the orientations indicate an understanding of the symmetry properties.*

5.17 ᘛ. Create so that the new figure has all three symmetry properties (vertical horizontal, 180° rotational).

5.18 ᏹ. Create so that the new figure has horizontal but not vertical symmetry.

5.19 ♉. Create so that the new figure has 180° rotational but not horizontal symmetry.

For the figures shown in Exercises 5.20–5.23 below, determine the smallest angle A such that the figure has A°-rotational symmetry (for a rotation about the center of the figure).

5.20 Colombian pottery, as shown in Figure 5.19.

5.21 Ceramic traditional American folk art, as shown in Figure 5.20. *Note*: Look carefully at the center pattern.

5.22 Pueblo pottery, as shown in Figure 5.21.

5.23 Basketry artwork, as shown in Figure 5.22. This is probably the bottom of a three-dimensional piece of artwork; separate patterns would appear around its sides.

For the strip patterns shown in Exercises 5.24–5.26 below, do the following:

(a) *Determine which symmetry properties are satisfied by the figure (vertical, horizontal, 180° rotational, glide reflect). You may assume the partial patterns continue as full patterns beyond the figure. As with the previous exercises, explain how the symmetry properties are satisfied by explaining how specific parts of the figure must by oriented by the artist in order to satisfy the property.*

(b) *If the overall pattern does not satisfy a particular symmetry property, there might be smaller elements of the pattern that do satisfy such properties. If*

Figure 5.19 Pottery from Colombia (culture not identified). (Courtesy of *Appleton's American Indian Designs*, Le Roy Appleton, Dover Publications, 2003, image 727.)

Figure 5.20 Ceramic traditional American folk art (Ohio, 1830). (Courtesy of *American Folk Art Designs*, Joseph D'Addetta, Dover Publications, 2006, image 130.)

Figure 5.21 Pueblo pottery. (Courtesy of *Appleton's American Indian Designs*, Le Roy Appleton, Dover Publications, 2003, image 426.)

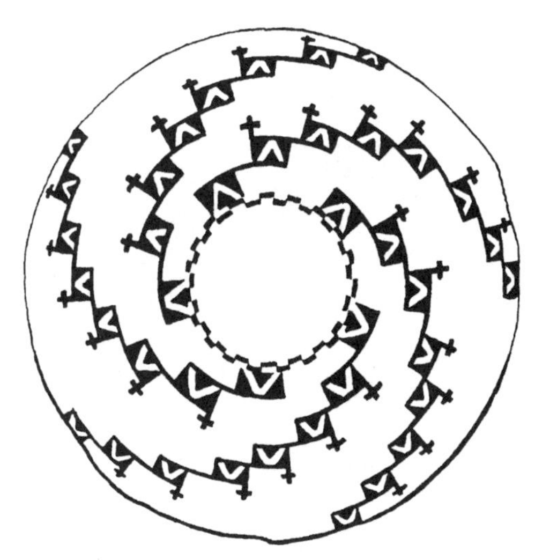

Figure 5.22 Basketry artwork of the Mission Indians, California. (Courtesy of *Appleton's American Indian Designs*, Le Roy Appleton, Dover Publications, 2003, image 373.)

that is the case, identify those parts that satisfy symmetry properties that are not satisfied by the overall pattern.

(c) *Determine the type (1 to 7) of the symmetry group to which the strip pattern belongs.*

(d) *Explain how the artist must understand the symmetry concepts by summarizing how the pattern must be created in order to satisfy the symmetry.*

5.24 French design, as shown in Figure 5.23.

5.25 Chinese design, as shown in Figure 5.24.

5.26 Persian design, as shown in Figure 5.25.

For the figures shown in Exercises 5.27–5.29 below, do the following:

(a) *Determine which symmetry properties are satisfied by the figure (vertical, horizontal, 180° rotational). As with the previous exercises, explain how the symmetry properties are satisfied by explaining how specific parts of the figure must by oriented by the artist in order to satisfy the property.*

Figure 5.23 French design from the 1700s. (Courtesy of *French Decorative Designs of the 18th Century*, Edouard Bajot, Dover Publications, 2006, image 204.)

Figure 5.24 Chinese design. (Courtesy of *Treasury of Chinese Designs*, Stanley Appelbaum (ed.), Dover Publications, 2006, image 073.)

Figure 5.25 Persian design. Stucco relief from the Haydariya madrasa in Qazwin, 11th to 12th century. (Courtesy of *Persian Designs and Motifs*, Ali Dowlatshahi, Dover Publications, 2007, image 213.)

(b) *Explain how color is used as part of the symmetry by explaining how specific parts of the figure are either the same color or an alternate color, with respect to the symmetry properties in (a).*

(c) *Determine the smallest angle A° such that the figure has A°-rotational symmetry (for a rotation about the center of the figure), as follows: First, determine the angle A° that gives back exactly the same figure, then determine the angle A° that gives back the same figure but with the colors alternated.*

(d) *Explain how the artist must understand the symmetry and alternating color concepts by summarizing how the pattern must be created in order to satisfy the symmetry.*

5.27 Artwork from a Turkoman carpet, as shown in Figure 5.26.

5.28 Celtic and Old Norse design, as shown in Figure 5.27.

5.29 Chinese design, as shown in Figure 5.28.

5.30 For each part below, create a set of patterns on a strip that satisfy the stated properties. You may use examples from this chapter, but it would be better to come up with your own design.

(a) Horizontal symmetry but not vertical symmetry.

(b) Glide reflect symmetry but not vertical symmetry.

Figure 5.26 Artwork from a Turkoman carpet. (Courtesy of *Persian Designs and Motifs*, Ali Dowlatshahi, Dover Publications, 2007, image 082.)

Figure 5.27 Celtic and Old Norse design. (Courtesy of *Celtic and Norse Designs*, Amy Lusebrink and Courtney Davis, Dover Publications, 2006, image 024.)

Figure 5.28 Chinese design. (Courtesy of *Treasury of Chinese Designs*, Stanley Appelbaum (ed.), Dover Publications, 2006, image 233.)

Figure 5.29 Artwork from the Hohokam culture of what is now the southwest United States. (Courtesy of *North American Indian Designs for Artists and Craftspeople*, Eva Wilson, Dover Publications, 1984, plate 5.)

(c) Vertical and 180° rotational symmetry and alternating colors.

(d) Vertical and 180° rotational symmetry and colors that do not alternate.

For Exercises 5.31 and 5.32, discuss the symmetry properties and three-dimensional artwork perspectives as we did regarding Figure 5.12.

5.31 Hohokam artwork, as shown in Figure 5.29.

5.32 Chiriguano-chané artwork, as shown in Figure 5.30.

Essay and Discussion

5.33 Most traditional artists would not describe their artwork as mathematical. Write/discuss how we can, nevertheless, conclude that the artist does indeed understand mathematical concepts.

5.34 Consider a culture such as the Navajo of the Southwestern United States, and notice how the patterns in Figure 5.31 of Navajo textiles (blankets) is "very symmetric" (it has all three of the symmetry properties of horizontal, vertical, and 180° rotation). The Navajo have a strong cultural belief in preserving harmony between humans and nature. Investigate the connection between Navajo beliefs in nature and how they are represented in Navajo art. Be sure to cite references.

5.35 Repeat Exercise 5.34 above for another culture with which you are familiar or can investigate.

Figure 5.30 Artwork from the Chiriguano-chané culture of what is now the Argentina. (Courtesy of *Argentine Indian Art*, Alejandro Eduardo Fiadone, Dover Publications, 1997, p. 72.)

5.36 Describe how you would design a machine that would automatically create cloth patterns that have various types of symmetry properties. If you can find resources on how such machines are made, or factories where such cloth products are made, that might give you some more information.

5.37 In some examples of art it appears that the figure "almost" satisfies a certain type of symmetry. That is, there is some small detail in the figure that does not satisfy the symmetry property, although the rest of the figure does satisfy that property. Write and/or discuss the reasons for this. In particular, what cultural explanations would there be for this minor detail?

5.38 Interview someone who creates textile art, such as quilts, weaving, embroidery, or knitting. Ask that person to describe how they create the patterns that appear in the final work. Ask that person to describe how much he or she knows about the pattern that will be used before beginning the project. Ask the person how long it has taken him or her to become skilled at making the art projects. If you cannot find someone who creates textile art, then look for someone who creates another kind of artwork, such as ceramic or wood, in which symmetry properties are a part of their work. Write/discuss your findings.

5.39 We now know that creating art often involves mathematical thinking. It turns out that in many traditional societies, the persons who create

Figure 5.31 Example of a Navajo blanket. (Courtesy of *North American Indian Designs for Artists and Craftspeople*, Eva Wilson, Dover Publications, 1984, plate 29.)

textile art are women. Investigate a culture in which most textile artists are women, and write/discuss the information you find on the following questions: How are textile artists treated in that culture? How are women treated in that culture? What kind of social status do experienced female artists have in that culture? If the culture was later controlled by another culture, did the role of women change? If so, how did it change? What do your findings indicate about how women who do mathematics are perceived in the culture you investigated? Compare what you learned about that culture with your own culture, in terms of how women who do mathematics are perceived.

Additional Comments

- In this chapter I have tried to include examples from many cultures around the globe. You may notice that there are not many examples from central Mexico nor from the Andean region of South America. This is because cultures from these areas, including examples of their artwork, will be presented in Chapters 9 and 10.
- At the risk of giving some preference to a particular source of material, I would like to point out that Dover Publications publishes many excellent collections of artwork from many parts of the world. If you wish to find more examples of artwork so as to study their mathematical properties, this would be one place you could find many such examples. On the other hand, some of their publications do not have complete details regarding the culture that made the artwork. In some cases there are excellent descriptions but in other cases the art is described only as being from some large area (e.g., "Africa" without specifying which part of Africa or which culture in Africa).
- It is also worth pointing out that one should always consider a variety of sources and, thus, look for examples from other places.
- I chose to discuss fewer details regarding symmetry and color so that I could devote more space to looking at a wide variety of examples. You will find more information about the symmetry groups and color in Ascher (1991: chapter 6) and the references therein.
- Another type of artwork that is very interesting to study is that of the *kolam* artwork created by women of Tamil Nadu culture of India. In Ascher (2002) you can find an excellent description of this case of cultural mathematics.

6

DIVINATION

Motivational Questions: Before you begin reading this chapter, take some time to think about your responses to the following questions.

- *Is there anything mathematical about trying to predict the future?*
- *How often do you think fortune-tellers are correct?*
- *Persons who do things like predicting the future or engaging in spiritual healing are usually seen as "unscientific," yet many people around the world seek the services of such persons. Why do you think that is?*
- *If you were to ask someone to predict your future, how many different types of outcomes do you think that would entail? Why?*
- *How many knots can you tie on a long blade of grass?*

INTRODUCTION

Imagine that we encounter a group of people from a culture about which we know little. They are sitting together and there are some unusual figures or objects between them. They start a process with one of the persons saying or chanting something to which the others pay close attention. Then the process continues with some of the objects being moved in special ways, and everyone

Introduction to Cultural Mathematics: With Case Studies in the Otomies and Incas, First Edition.
Thomas E. Gilsdorf.

looking closely at them. There may be cries of joy or of sadness. What is this group of people doing? Although there are many possibilities, two that we will investigate in this chapter and the next are: divination and playing games. Both of these processes could start in the way just described. In this chapter we will look at the cultural and mathematical aspects of divination, and in Chapter 7 we will look at the related activity of games. First, we will consider what kinds of activities are involved in divination. Then we will investigate two specific examples of divination, one from the Caroline Islands, the other from Madagascar. The cultural aspects of these activities will be part of our discussion throughout. Also, thinking back to Chapter 1 and Bishop's six mathematics categories, this chapter's topic is in the categories of location, number, and games ("games" here meaning some defined process, not necessarily a leisure process; more about this in the next chapter).

DIVINATION: GENERAL IDEAS

Divination is the process of predicting future events. There are interesting mathematical aspects that can be part of a divination process, but we need to know more precisely what goes on in a divination activity. So, first we must understand something about the general process of divination.

For the cases that interest us with respect to mathematical ideas, there are certain conditions that are common in most examples. Keep in mind that the goal of a divination event is to obtain an outcome that can be interpreted with regard to the person or persons seeking to know something about the future. With this in mind, what follows is the basic setup.

A process of divination typically starts with a person, a couple, or other close-knit group of people (I will call them *clients*) meeting with a person trained in identifying and interpreting possible future events (to whom I will refer as the *diviner*). Thus, the meeting would commence with a question from the clients to the diviner. It may have to do with wanting to know possible future events related to health, jobs, success in relationships, wealth, as well as other possibilities. Assuming the divination process is one that includes some mathematical process, then this means the diviner will go through some procedure or algorithm in order to select one or more possible outcomes, which I will call *destinies*. People lead complicated lives, so the collection of possible outcomes must necessarily be large enough to include quite a few potential outcomes. On the other hand, the general context of divination is that something about the future will be determined rather quickly, usually during the same meeting when the original question is posed to the diviner. We can infer from this that if a divination process involves such an enormous number of outcomes as to make the process of determining an outcome too long or too complicated, then that process would be impractical. Notice the connection between the social/cultural context of the divination process and the mathematical aspects, specifically the concept of number.

Here is another such connection. Virtually everyone, regardless of culture, would agree that having detailed knowledge of a person's future, with absolute certainty, is essentially impossible. Hence, if someone is curious to know about future events, there is a feeling that those future events depend on something beyond human control. In Western culture we refer to this feeling with terms such as "fate," "destiny", or "luck" (good or bad). Based on this assumption, a divination process that includes some mathematical activity will include a *random process* in it.

Finally, about the skill of the diviner, we can say the following. As with the topic of art and decoration, where the artists may keep track of large quantities of information and do many calculations mentally, a diviner also would often keep track of large quantities and do many calculations entirely from memory. We will see some specifics in a moment, but we can already deduce that the diviner must have accurate knowledge of the different types of possible outcomes of the divination process, which, as we have already seen, is necessarily a large number of possibilities. We now have enough background information to start looking at a specific example.

DIVINATION IN THE CAROLINE ISLANDS

The Caroline Islands are located in the South Pacific. Over the past several hundred years some or all of this collection of islands have been ruled by countries such as Spain, Japan, and the United States. Recently, they have gained independence as two separate countries. See "Caroline Islands" (2008) for more details. The people of the Caroline Islands have a process of divination that we will now discuss.

The divination process on the Islands has the following basic procedure. The client meets with the diviner and asks a question about something in the future. The diviner and the client tie varying numbers of knots in the leaves of a coconut tree. After a certain number of knotted leaves are made, they are put in a pile between the diviner and the client. The diviner picks up four of the knotted leaves in a random fashion and begins to count the number of knots in each one. As the diviner is counting the number of knots on a leaf, each time the number four is reached, the count starts over with one. This continues until all the knots on the leaf have been counted. This process is repeated until the knots on all four knotted leaves have been counted. As the knots on each leaf are being counted, the leaf is carefully held, and the order in which the four knotted leaves have been examined is kept track of as well.

Once the four knotted leaves have been examined, the diviner considers the outcome. The outcome is determined by the final numbers on each knotted leaf. The values are only between one and four because of the way the knots were counted, and each such value represents a destiny. In addition, the order in which the leaves were counted is also part of the process. Specifically, two pairs of numbers are constructed from the counting process: (a,b), and (c,d), where the order in which the knotted leaves were counted had value a for the first leaf, value b for the second, value c for the third, and value d for the fourth leaf.

We will continue the description of the interpretation of the information obtained in the divination process, but first we can make some mathematical conclusions. One is that the diviner counts in units of four. Instead of counting the knots as 1, 2, 3, 4, 5, 6, 7, 8, 9, 10, 11, and so on, the diviner cycles after four, like this: 1, 2, 3, 4, 1, 2, 3, 4, 1, 2, 3, and so on. In Western terms, we would say that the diviner understands *arithmetic modulo 4*. It is possible to count in this way for any integer (greater than one), and we will see in Chapter 8 how this kind of counting is done in the context of calendars. For our situation now of divination, let us put down the Western description of counting in blocks of numbers.

Definition 6.1 Two integers a and b are called *congruent modulo n* if their difference is a multiple of n. It is written like this: $a \equiv b \pmod{n}$ if $a - b = k \cdot n$ for some integer k.

An explanation is in order. First, all quantities are integers in the definition. Second, the integer n is assumed to be a positive integer and also $n > 1$. Third, the other values, a, b, and k, can be negative. (We will see situations with negative values in the context of calendars in Chapter 8.)

Now, let us examine how the Caroline Islands diviner's counting looks in this notation. If, for example, there are 15 knots on one of the leaves, then, recalling that $15 - 3 = 12$, we could say that $15 \equiv 3 \pmod{4}$. In fact, *any* positive integer (and even any negative integer or 0) is congruent to one and only one of the numbers 1, 2, 3, or 4. This means that no matter how many knots are tied on the leaves, it will always be possible to determine the values for the outcomes. The diviner is aware of this, considering that there are no numerical restrictions on the numbers of knots that can be tied on each of the leaves.

Moreover, the diviner understands the concept of an *ordered pair*, as we can see from the fact that the order matters in the process. Finally, the diviner understands the concept of a random process, which occurs when the knotted leaves are put in a pile and then four such leaves are randomly chosen. This connects us with the previous general discussion of divination in the sense that a random process is part of the process. Now, let us look at a few examples of the mathematics of this divination process. Keep in mind that the congruence modulo 4 notation is Western notation and that the diviner would not use this way of describing the quantities.

Example 6.1 Suppose the number of knots on the four selected knotted leaves comes out to be: 18, 13, 3, and 12. What are the ordered pairs that go with these values?

To determine the values, we calculate them modulo 4: $18 = 16 + 2$, so this means $18 \equiv 2 \pmod{4}$. Likewise, $13 = 12 + 1$, so $13 \equiv 1 \pmod{4}$. For 3, we simply have the value 3. For the last value 12, this number is already a multiple of four, so $12 \equiv 4 \pmod 4$. By the way, some people who work with modular arithmetic would write this last value as $12 \equiv 0 \pmod 4$; however, I will use $12 \equiv 4 \pmod 4$. Thus, the two ordered pairs are: (2, 1) and (3, 4).

Next, we continue with the cultural part of the divination process. The ordered pairs represent spirits that are connected with destinies. The first ordered pair identifies a specific destiny spirit about which the diviner knows by memory. The second ordered pair identifies a second destiny spirit in a similar manner. Once the two ordered pairs are determined, the diviner can describe the outcome to the client in terms of destiny spirits and the interpretation of them.

Now is a good time to calculate the total number of destinies that are possible in the Caroline Islands divination process. Recall that it should be large enough to cover a wide variety of life experiences, but not so large as to be unmanageable. We can use the reasoning of counting. That is, we count all possibilities, keeping in mind that each value occurs independently of the others and that the order is important. Because the diviner counts modulo 4, even though actual values may be larger, the final number of outcomes for each part of each ordered pair is four. Thus, there are four possibilities for the first value of the first ordered pair, four possibilities for the second value of the first ordered pair, and so on. The calculation looks like this:

$$(4,4) \quad (4,4),$$

which leads to

$$4 \times 4 \times 4 \times 4 = 256$$

total possible destinies.

ANOTHER DIVINATION PROCESS: SIKIDY

Our next example has to do with a divination process from Madagascar, though there are similarities between this process and divination processes in some other parts of the world. The name of the divination is *sikidy*. To start, let us understand some things about cultural tendencies in Madagascar. The main feature is that spatial orientation and direction are very important. The locations and positions of buildings in a typical village are also heavily influenced by factors of direction and spatial orientation. In particular, two lines, one from north to south and the other from east to west, are used as guidelines for positioning houses and rooms. The north-south line represents relationships between the living, and the east-west line represents relationships between the dead and the living. Hence, how buildings and rooms are located, positioned, and ordered with respect to these two lines reflects cultural beliefs and expectations about relationships within the community. Such beliefs and expectations include aspects of social status, known acquaintances versus strangers, moral values and expectations of different people in the community, and so on. In Ascher (2002: chapter 1), you can find specific details of cultural behavior and its relation to position and direction. What is interesting to us from the point of view of cultural mathematics is that we will see that the spatial orientation and direction aspects of culture in Madagascar are also expressed in the process of divination. In fact, it is time for us to discuss divination in Madagascar.

Recall that the overall process of divination has to do with the diviner asking the client questions about the situation and gathering information. In the context of the divination of *sikidy*, the diviner, called the *ombiasy*, draws conclusions regarding a fixed set of factors from the divination process. Specifically, there are sixteen considerations that comprise the factors. I will list them, using the letter *C* for considerations. We will see that the same letter will be useful for the mathematical notation of the sixteen factors. The considerations are:

C1: the client
C2: material goods
C3: an evil male entity
C4: the earth
C5: the child
C6: the bad intentions
C7: a woman
C8: the enemy
C9: the spirit
C10: nourishment
C11: the ancestors
C12: the road
C13: the diviner
C14: the people
C15: the creator
C16: the house.

To create a body of numerical information related to the divination, the *ombiasy* uses the concept of a random process to create such data. Our next step in the description of *sikidy* will be to describe that random process.

First, the *ombiasy* starts with a pile of seeds from a fano tree. The seeds are put into four piles by taking four handfuls of the seeds. Next, the ombiasy makes some incantations and indicates formal invitations to begin the divination event. The seeds are counted, but the final feature of the count has to do only with whether the count was even or odd. For our purposes, we will mark odd values by "o" and even values by "oo." As an example, consider that in each of the four handfuls, the number of seeds comes out to be: 17, 23, 14, and 19. This set of values would be denoted (note the order), as o, o, oo, o. This process continues for a total of four sets of four handfuls. The order of the values of each handful is very important, and we will use some Western notation, denoting the values as columns. There are a total of sixteen values, arranged in four columns of four. To see how these are arranged, we use an example. Let us now use the letter C again, this time to indicate columns that are formed by the ombiasy.

Example 6.2 Suppose the handfuls of seeds come out to be the values indicated below. You may wish to verify the patterns of o and oo that are obtained for columns 2, 3, and 4.

20, 31, 15, 16: oo o o oo: column 1 (one even number, then two odds, then one even).
21, 23, 29, 16: o o o oo: column 2.
20, 15, 30, 36: oo o oo oo: column 3.
31, 33, 29, 23: o o o o: column 4.

In our representation of how the *sikidy* process works, we arrange these four columns vertically, orientating them from *right to left*. Thus, we obtain the following, in which I will denote column 1 by C_1, column 2 by C_2, and so on:

C_4	C_3	C_2	C_1
o	oo	o	oo
o	o	o	o
o	oo	o	o
o	oo	oo	oo

Notice that column 1, which was of the form oo o o oo, is now vertical and at the far right of the display above. Now, more columns are created. The first set of columns is formed by denoting the horizontal values as columns. For instance, the top row of the above display is of the form o oo o oo, and this becomes column 5. So, columns 5–8 are formed as shown below:

C_4	C_3	C_2	C_1	
o	oo	o	oo	C_5
o	o	o	o	C_6
o	oo	o	o	C_7
o	oo	oo	oo	C_8

The table of values above is called the *mother sikidy*. From this array of information, all other pertinent aspects of *sikidy* are obtained. In fact, we will see shortly that there are eight more columns that are obtained. The final set of columns is formed by adding together certain of the previous columns. This means we will have to define what it means to add combinations of o and oo. I will describe the calculations of more columns, but first, there are some ideas about adding to which we must attend.

There is more than one way to define combinations of o and oo, and in Western mathematical terms defining such operations is part of what is called Boolean algebra, which is at the foundation of the study of circuit switching that is used in computers. In Ascher (2002: chapter 1) you can find an excellent description of details of Boolean algebra as it relates to *sikidy*. For our purposes, we can view the definitions of adding combinations of o and oo by keeping in mind that the notions of even and odd are the main properties under consideration in the process. From this point of view the operations can be deduced by thinking in terms of sums combinations of even and odd integers. Notice that because there are only two quantities to add (o and oo), we can construct the definition of all possible combinations by listing them out. The definitions of the operations are given below.

Definition 6.2 The sums of values of columns in the *sikidy* divination are as follows.

$$\begin{aligned} \mathrm{o} + \mathrm{o} &= \mathrm{oo} \\ \mathrm{o} + \mathrm{oo} &= \mathrm{o} \\ \mathrm{oo} + \mathrm{o} &= \mathrm{o} \\ \mathrm{oo} + \mathrm{oo} &= \mathrm{oo}. \end{aligned}$$

All of the main properties of real numbers are valid for this definition of addition with two symbols, including the associative and commutative properties. Moreover, recalling the properties of being a group, it can be proven (see Exercise 6.23) that two symbols with addition as defined here form a group. Recall the definition of group from Chapter 4, in our discussion of the Warlpiri kinship system. Can you guess which element (either o or oo) is the identity? Also, observe that for each symbol there is an inverse of that symbol.

We can define other operations with these symbols, one of those being *multiplication* (I will use the usual "×" symbol):

Definition 6.3 Multiplication of values of columns in the *sikidy* divination are as follows.

$$\begin{aligned} \mathrm{o} \times \mathrm{o} &= \mathrm{o} \\ \mathrm{o} \times \mathrm{oo} &= \mathrm{oo} \\ \mathrm{oo} \times \mathrm{o} &= \mathrm{oo} \\ \mathrm{oo} \times \mathrm{oo} &= \mathrm{oo}. \end{aligned}$$

Another operation is the *complement*, in which we switch from one symbol to the other (think on/off switch). In our context, using the notation ^, we have o^ = oo, and oo^ = o. Let's go through a couple of examples of the operations we have discussed so far.

Example 6.3 Evaluate the following expressions:

(a) (o + o) × (oo^).
(b) (o^ × oo) + o.

To evaluate these, we can proceed as we would for combinations of numbers; however, keep in mind the definitions of the various operations. So, (a) becomes (o + o) × (oo^) = (oo) × (o) = oo × o = oo. Likewise, (b) is evaluated as: (o^ × oo) + o = (oo × oo) + o = oo + o = o.

Example 6.4 Show that for any x and y (of o and oo), we have: $(x\text{^} \times y) + (x \times y\text{^}) = x + y$.

For this example, we can obtain the result by making a table of the possible outcomes (you can fill in details of some calculations as exercises).

x	y	$x+y$	$(x\text{^}\times y)+(x\times y\text{^})$	Equal?
oo	oo	oo	(oo^×oo)+(oo×oo^)=(oo)+(oo)=oo	yes
oo	o	o	(o)+(oo)=o	yes
o	oo	o	(oo)+(o)=o	yes
o	o	oo	(oo)+(oo)=oo	yes

With these examples, we can now go back to our example from *sikidy*. The following is how eight more columns are calculated:

$$
\begin{aligned}
C_9 &= C_8 + C_7 \\
C_{10} &= C_6 + C_5 \\
C_{11} &= C_4 + C_3 \\
C_{12} &= C_2 + C_1 \\
C_{13} &= C_9 + C_{10} \\
C_{14} &= C_{11} + C_{12} \\
C_{15} &= C_{13} + C_{14} \\
C_{16} &= C_{15} + C_1.
\end{aligned}
$$

At this point, the *ombiasy* has the sixteen considerations, now represented by sixteen columns. The *ombiasy* arranges all sixteen columns, but the arrangement is not arbitrary. The columns are put in locations that are connected with how each one was formed. For example, the first four columns are at the top, indicating the first block of data. In fact, at this moment, it is important for us to determine how many destinies, that is, outcomes, are possible for this divination process. Thinking back to the beginning of this chapter. We want to have enough outcomes to account for the complexities of people's lives, but not so many that would be too difficult to manage. In the case of *sikidy*, we have sixteen columns, each of which has four entries for which there are two possible values. At first it may seem that we would have $(2^4)^{16} = 1.84 \times 10^{19}$ destinies! This would be too many possible values to deal with; however, the reality is that there are not this many possibilities. Note that the first four columns are *only* the ones that contain randomly generated data. All other columns are formed by some kind of relation to columns that were previously formed. Thus, the number of outcomes, that is, destinies, is calculated by four columns, each having four entries, with two possible values for each. Each column consists of four places to put possible values, and for each place there are two choices. Hence, there are 2^4 possible outcomes for each column. Because the values obtained in each entry of each column and between columns are all independent of each other, the total count consists of the product of all possible outcomes, as shown below.

$$2^4 \times 2^4 \times 2^4 \times 2^4 = (2^4)^4 = 65{,}536.$$

This is the number of destinies. Now we will continue Example 6.3, by calculating some of the rest of the columns. It is worth noting that the calculations, just like the alignment of the first four columns, is done from *right to left*. Here are the details. We had left off with the mother sikidy as:

C_4	C_3	C_2	C_1	
o	oo	o	oo	C_5
o	o	o	o	C_6
o	oo	o	o	C_7
o	oo	oo	oo	C_8.

To obtain more columns, here is how we add them:

$C_9 = C_8 + C_7$: Add:

o	oo	o	o	C_7
o	oo	oo	oo	C_8,

obtaining C_9 as:

$$C_9 = \begin{array}{c} \text{o} \\ \text{o} \\ \text{oo} \\ \text{o} \end{array} + \begin{array}{c} \text{oo} \\ \text{oo} \\ \text{oo} \\ \text{o} \end{array} = \begin{array}{c} \text{o} \\ \text{o} \\ \text{oo} \\ \text{oo} \end{array},$$

where the top entry of C_9 is obtained by adding the far right entries of C_7 and C_8, that is, o + oo = o, then the second entry of C_9 is obtained by adding the second-from-the-right entries of C_7 and C_8, that is, o + oo = o (again), and so on. As another example, let us calculate $C_{11} = C_4 + C_3$:

$$C_{11} = \begin{array}{c} \text{o} \\ \text{o} \\ \text{o} \\ \text{o} \end{array} + \begin{array}{c} \text{oo} \\ \text{o} \\ \text{oo} \\ \text{oo} \end{array} = \begin{array}{c} \text{o} \\ \text{oo} \\ \text{o} \\ \text{o} \end{array}.$$

The next step is to arrange all sixteen columns. The general diagram is shown in Figure 6.1.

We already knew the basic arrangement of the first eight columns, but columns nine through sixteen do not seem to be ordered (either right to left or left to right). Why is that? The answer is that we can remind ourselves that the divination is very much a *cultural* event, and as such, factors such as relationships, orientation, and relative position are important, more so than numerical ordering. Thus, the columns are arranged in a way that reflects their inter-relationships. For example, column thirteen is between column nine and

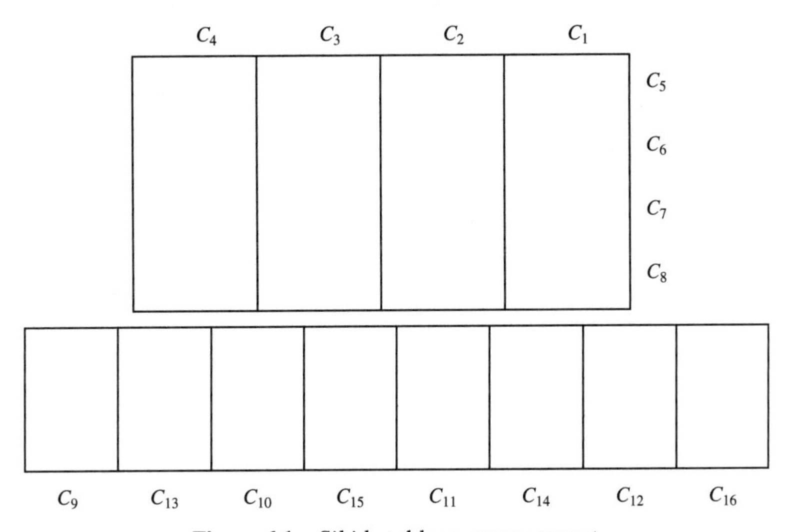

Figure 6.1 *Sikidy* tableau arrangement.

C_4	C_3	C_2	C_1	
o	oo	o	oo	$\leftarrow C_5$
o	o	o	o	$\leftarrow C_6$
o	oo	o	o	$\leftarrow C_7$
o	oo	oo	oo	$\leftarrow C_8$

o	oo	o	oo	o	oo	o	oo
o	o	oo	o	oo	oo	oo	oo
oo	o	o	oo	o	o	oo	o
oo	oo	oo	o	o	o	oo	o
C_9	C_{13}	C_{10}	C_{15}	C_{11}	C_{14}	C_{12}	C_{16}

Figure 6.2 Final *sikidy* tableau arrangement for Example 6.2.

column ten because $C_{13} = C_9 + C_{10}$. Figure 6.2 shows the final set of columns for our example that we started some time ago with Example 6.2. In the exercises, you can verify the column values that we have not already calculated.

The array of sixteen columns is called the *final tableau*. Once the *ombiasy* has these sixteen considerations, which now are represented as sixteen columns of numerical information, the *ombiasy* can finish the divination process. Before continuing with the divination process, however, the *ombiasy* checks several mathematical properties of the sixteen columns. For example, the sum of the entries in C_{15} must be oo, that is, an even number. Let us check that for our final tableau: Adding the elements of our C_{15}, we obtain (adding from top to bottom): oo + o + oo + o = oo: two odds plus two evens, being an even, gives us oo, as predicted.

Well enough, but how does the *ombiasy* know that the sum of the elements of C_{15} will be even for *any* final tableau? Let us try to determine that for ourselves. By its definition,

$$C_{15} = C_{13} + C_{14} = (C_9 + C_{10}) + (C_{11} + C_{12}),$$

and by continuing with the definitions, we replace C_9, C_{10}, C_{11}, and C_{12} with the definitions of their values to obtain:

$$C_{15} = C_8 + C_7 + C_6 + C_5 + C_4 + C_3 + C_2 + C_1.$$

Let us try to understand what the above statement implies. Recall that columns five through eight consist of the same values as those in columns one through four; it is just that the values are read horizontally instead of vertically. What C_{15} represents is the sum of all columns one through eight. This in turn represents the values of the first four columns, then the values of those same four columns, added together, for a total of *twice* all the values of the first four columns. Regardless of whether the sum of the first four columns is even or odd, the double of that sum, which is exactly what C_{15} represents, must be an even number. Notice that it does not matter what the values of the four columns are, and so the result is valid for any final tableau. Let us look at another mathematical property that the *ombiasy* checks.

In any final tableau, there are always at least two columns that are equal. To see how this can be, let us calculate values. First, recall from our discussion of the number of possibilities that there are $(2^4)^4 = 65{,}536$ possibilities for any final tableau. In fact, we can think of this value as $16 \times 16 \times 16 \times 16 = 65{,}536$. Now, to examine the claim that at least two columns must be equal, observe that if every column in the final tableau were distinct, then we would have sixteen columns of all distinct values, and that would represent all possible combinations of the values of the columns. It turns out that we can determine the sum of all sixteen columns: Because the sixteen columns would represent all possible combinations, that sum would consist of sixteen columns in which we vary all possible combinations of o and oo (the combinations come in pairs). This sum must necessarily be even. Hence, the sum of all sixteen columns would be a column that consists of all oo entries. Let us refer to this column by C, that is,

$$C = \begin{matrix} \text{oo} \\ \text{oo} \\ \text{oo} \\ \text{oo} \end{matrix}.$$

Another observation is that if we are assuming that all possible combinations of o and oo are represented in the sixteen columns, then one of those columns must be one that has all entries of oo. This will be useful information shortly. Next, we can add all of the columns together again, noting some of the relationships between the columns that are formed as sums of other columns. Our first sum is of *all* sixteen columns. Here we go:

$$C = C_1 + C_2 + C_3 + \cdots + C_{16}.$$

We can rewrite this sum, recalling that the first eight terms add to C_{15}. Thus,

$$\begin{aligned} C &= C_8 + C_7 + C_6 + C_5 + C_4 + C_3 + C_2 + C_1 + C_9 + C_{10} + \cdots + C_{16} \\ &= C_{15} + C_9 + C_{10} + \cdots + C_{16}. \end{aligned}$$

Now let us apply the definitions of the sums of the remaining columns. In particular, we replace $C_9 + C_{10}$ by C_{13}, and replace $C_{11} + C_{12}$ by C_{14}.

$$\begin{aligned} C &= C_{15} + C_9 + C_{10} + C_{11} + C_{12} + C_{13} + C_{14} + C_{15} + C_{16} \\ &= C_{15} + C_{13} + C_{14} + C_{13} + C_{14} + C_{15} + C_{16}. \end{aligned}$$

In Exercise 6.16, you can see that for any column C_i, we have $2 \times C_i = C$, and $C + C_i = C_i$. In other words, twice any column gives us all entries of oo, and C acts like zero, leaving any other column unchanged under addition. Applying this to our sum of the sixteen columns, we arrive at:

$$C = 2 \times [C_{15} + C_{13} + C_{14}] + C_{16} = C + C_{16} = C_{16}.$$

Meanwhile, by definition, $C_{16} = C_1 + C_{15}$, so $C = C_1 + C_{15}$. By Exercise 6.16b, we conclude that $C_{15} = C_1$. Thus, there are two equal columns. What we have done, using Western mathematical terms is to prove that two columns must be equal by *contradiction*. That is, we assumed that all of the columns were distinct, yet went on to show that $C_{15} = C_1$. Our original assumption that all columns are distinct must be false, so we conclude that two (at least) must be equal. Notice also that it was column C_{16} that came out to be the column of all oo.

Let us check to see if this holds for our final tableau of Example 6.2, shown in Figure 6.2. Looking at it carefully, you should notice that columns one and thirteen are equal, and there is another pair of equal columns (can you find them?). Wait a minute: Columns one and fifteen are not equal! Is there something wrong? To resolve this question, we can think back to our process, which included the assumption that all sixteen columns must initially be distinct. In practice, we now know that this will not happen. In fact, it could happen that two or more of the first four columns could be equal, by similar combinations of even and odd numbers of seeds. Because we assumed that all possible columns of four values of even or odd numbers are present, we know that it must be that one of those columns must be C, the column of all evens (oo). In the discussion above, we calculated the sum of columns in a rather particular way, at one point having $C_{16} = C$. We could do the proof again, this time assuming that a column other than C_{16} is the one equal to C. In that case, we obtain two equal columns, but they are not columns C_{15} and C_1. Exercise 6.20 asks you to go through a specific procedure such as this, in which the equal columns are different from C_{15} and C_1. Our next item is to discuss the *cultural aspects* of the *sikidy* divination process.

Apart from the relative location of the columns, there are spatial orientation aspects of the process as well. We have seen that the position of the columns is important because of the relations between the columns, such as through addition of previous columns. Also, the *ombiasy* pays close attention to direction by examining lines through the columns and the values of the columns with respect to those lines. For example, a line representing a path from the southwest to the northeast is formed in the tableau, with some regions representing an area referred to as the land of the slaves, while other areas represent an area referred to as the land of princes. There are migrators that move from one area to another as well. The possible relationships between slaves and princes is part of the divination process that the *ombiasy* interprets. Recall from the beginning of this section that spatial orientation and direction are very important in the culture of Madagascar. We can see that the mathematical information formed during the *sikidy* divination process is also an expression of this importance of spatial orientation and direction. Notice the wonderful connection between mathematics and culture!

Many more calculations are possible. The *ombiasy* applies as many 100 different formulas to the final tableau! See Ascher (2002: chapter 1).

FINAL REMARKS

We should be careful not to underestimate the overall skill of a diviner, apart from that person's mathematical skills. Recall that the clients come to the diviner to ask questions about their personal lives. In many such contexts, the diviner might ask the clients to elaborate on their questions, and such elaboration would provide further information to the diviner about the clients' situations. Moreover, an experienced diviner who has encountered numerous types of settings in which clients have asked questions inevitably gains insights into human, cultural, and other nuances that are relevant to what may or may not happen in the future. In short, despite the fact that a divination process may at first appear to be superficial, the diviner is, in reality, very skilled. Finally, the diviner is typically a respected person within the community or culture. As Ascher describes regarding the Caroline Island diviners, "The teachings of the spirits are known to the diviners, who are sacred and honored people. Not only are the diviners consulted on most important matters, including fishing, house building, traveling, naming of children, illness, and love, but they must carefully pass on their knowledge by teaching future diviners" (Ascher, 2002: 8).

EXERCISES

Short Answer

6.1 Are there any activities in mainstream Western culture that resemble divination processes? If so, describe them.

6.2 True or false: The divination scheme of the Caroline Islands can be considered part of the history of the United States. Explain your response.

6.3 For each of the following descriptions, explain why the setup would not lead to a good process of divination.

(a) Two values are obtained, each of which could have three possible outcomes.

(b) Twenty values are obtained, each of which could have seven possible outcomes.

6.4 Explain the mathematical concepts that a Caroline Islands diviner understands as part of the divination process. Think of as many as you can.

6.5 Explain the mathematical concepts that a *sikidy* diviner understands as part of the divination process. Think of as many as you can.

6.6 **(a)** Is it possible for a final tableau to have *four* equal columns? Explain.
(b) Is it possible for a final tableau to have *six* equal columns? Explain.
(c) Is it possible for a final tableau to have *all* equal columns? Explain.

Calculations

6.7 For each of the knot values below, determine the ordered pairs of the Caroline Islands divination process. Recall that order counts.
(a) 31, 18, 16, 1 **(b)** 13, 11, 5, 17.

6.8 **(a)** Suppose the diviner selects four strips that contain the knot values (in the following order): 13, 6, 19, 9. Determine the two ordered pairs that result from this outcome as in Exercise 6.7 above.
(b) If the knot values had come out to be 9, 19, 6, 13, would the client's fortunes be better or worse? Explain your response.

6.9 Repeat Exercise 6.8 above for the following values of numbers of knots (recall that order counts): 22, 2, 22, 1.

6.10 Suppose we create a variant of the Caroline Islands divination by replacing the counting by fours aspect to counting by sixes.
(a) Determine the total number of possible destinies for this new divination process.
For parts (b) and (c) below, determine the ordered pairs of this new divination process if the number of knots are as given.
(b) 23, 33, 16, 3. **(c)** 46, 19, 29, 35.

6.11 Suppose for a process of divination, the diviner first randomly selects two of the following symbols: %, &, @, #, *, A, B, C, D, E (repetitions are possible). Then, the diviner creates an ordered pair from the two selected symbols. Next, the diviner randomly selects two of the following symbols: Δ, β, and ∂ (repetitions are possible), and creates a second ordered pair. The final result depends on the two ordered pairs created from each of these steps.
(a) Determine the number of destinies that are possible for this process of divination.
(b) What does the total number of possible outcomes (destines) tell you about whether this would be a reasonable process for divination? Explain.

6.12 Fill in details of Example 6.4, for the remaining three cases.

6.13 Verify the details of the calculations of C_{10}, C_{12}, C_{13}, C_{14}, C_{15}, C_{16} of the final tableau shown in Figure 6.2.

6.14 Suppose the *ombiasy* obtains the following values for the counts for the fano seeds (in the order given): First: 18, 21, 19, 30; second: 22, 34, 13, 17; third: 15, 22, 11, 9; fourth: 12, 8, 16, 31. Use this information to set up the *mother sikidy*, and then construct the final tableau from it. *Note*: This will take some time, but you get to go through the entire process. Also, remember to line up the columns in the manner shown in Figure 6.1.

6.15 Refer to your tableau from Exercise 6.14 above.

(a) Determine which two columns are equal.

(b) Note that they are *not* columns 1 and 15, as obtained in the proof that at least two columns must equal. Explain why this does not contradict our proof that two columns must be equal in any final tableau.

6.16 **(a)** Use the definition of addition of columns in the *sikidy* divination process to verify that for any column C_i, we have $C_i + C_i = 2 \times C_i$. *Note*: The method is to check the different combinations of values that could occur in each side of the expression.

(b) Similarly, prove that if C_i and C_j are any two columns of a *sikidy* tableau such that

$$C_i + C_j = \begin{matrix} \text{oo} \\ \text{oo} \\ \text{oo} \\ \text{oo} \end{matrix},$$

then it must be that $C_i = C_j$.

(c) Is the result of part (b) true for three columns? That is, if

$$C_i + C_j + C_k = \begin{matrix} \text{oo} \\ \text{oo} \\ \text{oo} \\ \text{oo} \end{matrix},$$

does it follow that $C_i = C_j = C_k$? Either prove that it is true or find three distinct columns that add like this but are not equal.

6.17 Use the definition of addition of columns in the *sikidy* divination process to verify that for any x and y, it must be that $x + y = (x + y) \times (x \times y)^{\wedge}$. *Note*: The method is like that in Example 6.4: Check the different combinations of values that could occur between x and y, then calculate each side of the proposed equation to see if they are equal.

6.18 For sixteen columns of two possible values (o or oo), there are a total of $(2^4)^{16}$ possible outcomes, but this is not the number of destinies that are possible from the *sikidy* process. Determine the correct number of destinies (i.e., outcomes) of the *sikidy* process. *Hint*: Think about the random part of the process.

6.19 Suppose that a *sikidy*-type divination is created in which there are three initial columns of randomly generated values of o and oo (instead of four). Suppose that from the three initial columns, more are created by various combinations (using +) of the first three for a total of twelve columns. Calculate the number of destinies of this divination process. Does the combination of the other nine columns have an effect on the number of destinies? Explain.

6.20 Suppose that in the proof of equality of at least two columns of a final tableau of *sikidy* divination, we assume that C_2 is the column of all entries oo (i.e., assume $C_2 = C$).

(a) Calculate the sum of all sixteen columns as we did in that proof, this time replacing columns using the following: $C_9 = C_8 + C_7$, $C_{10} = C_6 + C_5$, $C_{11} = C_4 + C_3$, and $C_{16} = C_1 + C_{15}$. Then obtain the equation $C = 2 \times [C_{11} + C_{10} + C_9 + C_{16}] + C_2 + C_{12} + C_{15}$. Use this equation and the ideas of the original proof to conclude that $C_{12} = C_{15}$.

(b) Explain how to modify part (a) so that we would conclude that $C_2 = C_{12}$.

Exercises 6.21–6.24 assume some experience with writing mathematical proofs and abstract algebra.

6.21* Here are two properties of modulo 4 arithmetic that the Caroline Islands diviners might know:

(a) Prove that if b is any *even* integer, then $b^2 \equiv 0 \pmod 4$.

(b) Prove that if b is any *odd* integer, then $b^2 \equiv 1 \pmod 4$.

6.22* Here is another property of modulo 4 arithmetic that the Caroline Islands diviners might know: Prove that if b is any *odd* integer, then $b^2 \equiv 1 \pmod 8$.

6.23* **(a)** Prove that the set {o, oo}, with the binary operation of + as defined by the addition of columns of the *sikidy* divination process, is a group. Is this group abelian? (Recall from abstract algebra that a group with an operation * is *abelian* if for all elements a and b, one has $b * a = a * b$; the operation is *nonabelian* if the equation $b * a = a * b$ does not hold for all elements a and b.) Explain.

(b) Consider the set {o, oo}, with the binary operations of + and × as defined by the addition and multiplication of columns of the *sikidy* divination process. Is this an integral domain? A ring? A field? Prove or disprove your responses.

6.24* Consider the proof in this chapter that at least two columns of the final tableau the *sikidy* must be equal. Construct another proof of this fact that is distinct from the proof presented in this chapter, and is also distinct from the proof outlined in Exercise 6.20.

Essay and Discussion

6.25 Create your own Caroline Islands type of divination scheme. You may wish to (randomly) draw knots on line segments (rather than tie knots on real plants). You will have to decide such things as how many segments (tied "leaves") are part of the process, how the knots will be counted (it should be a number other than four or six, which have already been considered), what the ordered pairs will mean regarding the response to the person asking the question, and how to determine what kinds of ordered pairs denote which kinds of destines. Experiment in small groups with the process and write/discuss your findings.

6.26 Write about or discuss how we can conclude that the first people to use the *sikidy* process must have developed a *proof* of the property of two equal columns. Consider such factors as the number of possible outcomes and the reliability of the process.

6.27 Write about or discuss a divination process other than those we have seen from the Caroline Islands and *sikidy*, discussed here, such as *geomancy*, *ilm al-raml*, *Ifa*, or *Bamana*. Explain the cultural background of the people who practice the divination process, the mathematical ideas involved, and the connection between the mathematical part of the process and the cultural part of interpretation of the divination.

Additional Comments

- Most of my writing of this chapter has been motivated by Ascher (2002: chapter 1). I have discussed some parts of the topic of divination in less detail than Ascher's work, so readers who would like to dig deeper into the cultural mathematics of divination will find Ascher's work an excellent resource.
- Related to the Caroline Islands divination. For readers with some experience with abstract algebra or number theory: The description of why every positive integer must be congruent modulo 4 to exactly one of the values 0, 1, 2, 3 comes from the fact that congruence modulo n ($n = 4$ here) is an equivalence relation that separates the set of all integers into

equivalence classes. In our case, there are four such equivalence classes, namely, the four that contain the values of 0, 1, 2, and 3. A characteristic of equivalence classes is that every element of the set (in this case the integers) belongs to one and only one equivalence class. We saw the concept of equivalence classes in Chapter 4 on kinship, so you can see that this concept appears in a variety of contexts.

- There are many types of divination schemes. Exercise 6.27 lists a few possibilities, and we will look briefly at divination of the Otomies in Chapter 9. Some divination schemes are similar in nature, a fact that was pointed out to me by Iman Chahine (personal communication).
- Medical anthropologists have devoted much effort to studying activities such as spiritual healing and have found that there are many cases in which such healing processes can have positive effects, even when Western medicine does not. See Womack (2010: chapter 11). We have seen in this chapter that diviners are not simply making wild guesses about the future. Their processes can have specific (mathematical) structure, and they incorporate other social tools such as psychology and social work into their processes. A diviner often plays an important role in the community, as a source of information and guidance. For these reasons, people in many cultures, including from Western culture, continue to seek the services of diviners, spiritual healers, and so on.

7

GAMES

Motivational Questions: Before you begin reading this chapter, take some time to think about your responses to the following questions.

- *What is a game?*
- *What is the difference between a card game and a game like chess or checkers?*
- *Why do people play games?*
- *If there are three choices for the main meal on a menu and two choices for a side item, how many total choices are there?*
- *What is the social context of playing games? How important are they in your life?*
- *The Western mathematical specialty called graph theory involves concepts connected with geometric shapes that have certain numbers of paths emanating from each vertex or intersection. Do you think this has anything to do with strategies involved in board games?*

INTRODUCTION

Recall from the last chapter on divination that we had an imaginary encounter with a group of people gathered together, engaged in a special process. In that

Introduction to Cultural Mathematics: With Case Studies in the Otomies and Incas, First Edition.
Thomas E. Gilsdorf.

chapter the process we studied was that of divination. In this chapter we will look at games. You will probably notice some similarities, such as cultural importance of both activities. On the other hand, we will see that games have more to do with probabilities than with outcomes such as those in divination. First, we will look at some general ideas related to games, followed by some analyses of specific games. Specifically, we will look at a game of logic of the Kpelle culture of Africa, then we will look at the board game of Hat Diviyan Keliya from Sri Lanka. Finally, we will consider a dice game, Pay-Gay-Say, of the Ojibwe culture of North America. To understand the Ojibwe game we will need to understand some ideas from the discipline of probability as well. As with virtually every topic of this book, the cultural connections of these activities will be part of our discussion throughout. Also, although this chapter is in the category of games, thinking back to Chapter 1 and Bishop's six mathematics categories, it is important to realize that "games" involves much more than a leisure activity. We will see this throughout the chapter.

GAMES: GENERAL IDEAS

Games can be found in many cultures, and go back a long time in human history. For the purposes of our discussion, I will organize games into four broad categories:

- Those that depend primarily on physical skill;
- Those that depend primarily on strategy or mental skill;
- Those that depend primarily on probability (chance, luck);
- Combinations of the previous three types.

Let us look at some examples of different types of games. An example of a game that requires mainly physical skill would be a short foot race. Here, the people involved run a short distance (maybe ten to fifteen meters), starting at a standstill (i.e., without any special equipment like starting blocks). The winner would be the person who arrives first at the finish line. Although there might be some strategy involved in anticipating the signal to start, most of this game depends on the physical skill of running as fast as you can. Notice that if we change the distance involved to, say, one kilometer, then the game involves both physical skill and strategy, because it is not possible for a person to simply run as fast as he or she can for that kind of distance. Chess is a game that involves mainly strategy and mental skill. The outcome of the game depends entirely on the strategy and mental skill of the players, with essentially no physical skill involved in the process.

For a game involving probability, that is, chance, I first must clarify that the meaning I intend with the word "chance" is that there is some random process that is part of the structure or rules that are in effect during most of the game.

The kinds of random processes that we would encounter in the context of games would come about by, for example, rolling a die or dice, tossing a coin, using a spinner (in the sense of a pointer on a circular pattern), or drawing a card from shuffled deck. Thus, for example, wind on a golf course is not considered part of chance in our discussion. Indeed, although one could say that gusts of wind represent essentially an unpredictable event, the aspect of wind is not an intentionally random process that is part of the rules of golf. Moreover, an astute player who observes gusts of wind during a game can alter how to play in order to compensate for the wind, hence using strategy. Similarly, a coin toss at the beginning of a game is a random event; however, this random process only occurs once and is not utilized during the main process of the game. Wind gusts in golf and coin tosses to start a game would not put the game under consideration into the category of having chance as a main part of its activity. An example of a game that involves only chance could be a game in which six people each choose a number (integer) between one and six, and then a die is rolled with the winners being the persons whose chosen number coincides with the number that shows on the die. Assuming the die is fair (i.e., all numbers from one to six are equally likely as outcomes), the game depends only on chance.

Finally, for a combination of types, we can consider a card game in which players draw from a shuffled deck and try to form certain combinations such as three or four consecutive cards of the same suit or three or four cards of the same type (e.g., three or four queens). In such games the winner is often the person with no cards left. For this kind of game, the random aspect comes from the draw of the cards from the shuffled deck, and the strategy comes from deciding how to play the collection of cards that has been drawn. For example, if you have a ten of spades and a ten of diamonds in your hand plus an eight and nine of hearts, then if you draw a ten of hearts from the deck, which combination do you make? There are two possibilities, that of forming three tens and that of forming the consecutive combination of an eight, nine, and ten of hearts. Your decision depends on the rules of the game and what the other players have played so far, such as the kinds of cards that have already been put down. So, strategy is involved.

On the cultural side of this topic, games often have important implications. For example, for a young child, playing games may seem to be an idle pastime; however, the child is in fact learning important thinking and physical skills during the process. In this chapter we will look at some examples of games and their cultural aspects.

A LOGICAL GAME

Our first example falls under the second category described above. That is, it is a game of mental skill. It is a popular riddle that requires logical reasoning to solve. Typically, it is a riddle involving three objects that must be transported

TABLE 7.1 A Solution to the Kpelle Game of Example 7.1

Step	Left Behind	Crossing the River	On the Other Side
1	Leopard, cassava	Man and goat →	
2	Leopard, cassava	← Man	Goat
3	Leopard	Man and cassava →	Goat
4	Leopard	← Man and goat	Cassava
5	Goat	Man and leopard →	Cassava
6	Goat	← Man	Cassava, leopard
7		Man and goat →	Cassava, leopard
8			Man, cassava, leopard, goat

across a river in such a way that certain combinations are not allowed. My version of this riddle comes from Zaslavsky (1999).

Example 7.1 The Kpelle of Africa tell the following version of the riddle: A man has a leopard, a goat, and a pile of cassava leaves to bring across a river. How can he take them across the river, without the goat eating the leaves and the leopard eating the goat?

Note that there are some assumptions being tacitly made in this problem: One is that a leopard will not eat the cassava leaves. Another is that the man can transport only one thing at a time. The other assumptions are that any other of combination is not allowable, specifically the leopard and goat cannot be together and the goat and the cassava leaves cannot be together. One solution to this problem has the following steps, shown in Table 7.1.

Other observations we can make about this game is that it is put in the setting of objects that are likely to be part of the local culture in which the game appears. In other parts of the world where this same riddle appears, the objects change. See Ascher (1991: chapter 4). Another observation is that the assumptions of this particular riddle can be modified so as to make the solution different, the problem more difficult, and so on. In the exercises, you will see a few of these possibilities.

MORE LEOPARDS, AND TIGERS! A GAME FROM SRI LANKA

There are many board games that have themes of one player pursuing another, or each trying to corner the other. In parts of Asia such games often involve tigers and leopards, with variations by region, of course. We will now look at one of these games, called Hat Diviyan Keliya, "the seven leopards," from Sri Lanka. Years ago, Sri Lanka was called Ceylon. You may wish to look

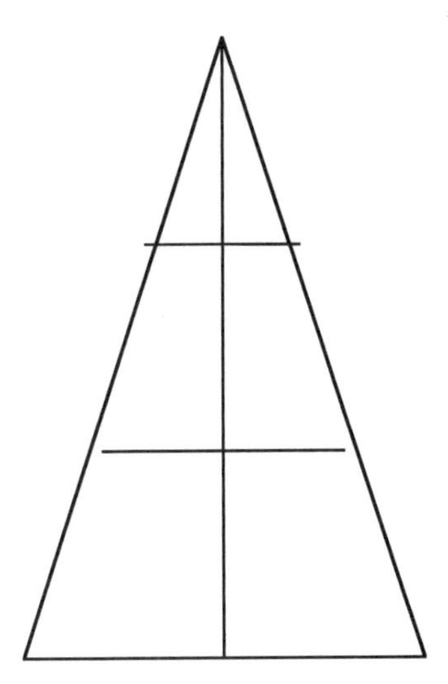

Figure 7.1 The game board for Hat Diviyan Keliya.

Sri Lanka up on a world map (it is near India). Now for the details of the game.

The game board consists of an isosceles triangle with one vertical line along the height and two horizontal line segments drawn partway down the vertical length of the triangle, as shown in Figure 7.1. One player has one playing piece that represents a tiger. The other player has seven pieces that represent leopards. The object of the game is for the tiger to try to eliminate leopards by jumping over each one to an empty intersection or vertex, while the leopards try to corner the tiger by positioning themselves so that the tiger has no empty place to which it can move and cannot jump any leopard. The player with the tiger wins if so few leopards are left that it is not possible for the leopards to corner the tiger. The player with the leopards wins if the tiger can no longer make any moves and cannot jump any leopards. All game pieces are on either a vertex or an intersection of the game board, which, as you can see, has ten such points.

The playing of the game proceeds as follows. The player with the tiger puts the tiger at the vertex at the top of the game board (i.e., at the apex of the triangle), then the other player begins placing the seven leopards anywhere else on the game board. In some versions no player makes any moves until all pieces are placed. In other versions the tiger is first placed at the top and can move each time the other player places a leopard on the game board. In this second version, the leopards cannot move until all seven have been placed on the game board. The strategy changes, depending on who can move when. To experience some mathematics of this game, let us suppose that one way or another all eight pieces are on the board. As the game

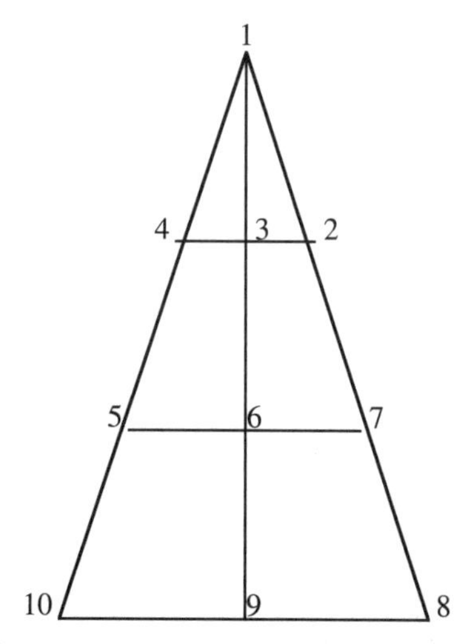

Figure 7.2 The labeled game board for Hat Diviyan Keliya.

progresses and some leopards are eliminated, we can ask ourselves about the possible strategies the player with the leopards can employ and how they are connected with the various points on the game board. Example 7.2 will be about that kind of situation; however, before looking at the example, let us label the ten points on the game board with numbers one through ten, so that we can easily refer to each one. The labeled game board is shown in Figure 7.2.

Example 7.2 Is it possible to corner the tiger at the point marked 7 if there are four leopards remaining? Give a careful explanation.

To trap the tiger at point 7, there would need to be a leopard at point 6, for instance. However, so that the tiger would be unable to jump the leopard at point 6, there would need to be another leopard at point 5. Similarly, we would need two more leopards, at points 2 and 1. The four leopards have now been accounted for, but point 8 remains open and the tiger could move there. It is not possible to have any of the four leopards at point 8 because that would mean there would be an open a point where the tiger could either move or jump. Conclusion: It is *not* possible to corner the tiger at point 7 with four leopards. Of course, perhaps you are now wondering if the player with the leopards must necessarily lose when there are four leopards left. Exercise 7.7 asks you to figure out the answer to this question. We can ask ourselves a question about the strategy from the tiger's point of view, in the next example.

Example 7.3 Which of the points on the game board are most advantageous for the tiger? Give a careful explanation.

Thinking about the previous example, you can observe that the more paths the tiger has to move toward, the less likely it is to be cornered by the leopards. Points 3 and 6 each have four paths emanating from them, so these would be the points at which the tiger would be less likely to get cornered by the leopards. Is this the *best* point for the tiger? You can check that, at either point 3 or point 6, there would need to be at least five leopards in order to corner the tiger at one of these points. On the other hand, at point 1, it would take at least *six* leopards to corner the tiger. Can you figure out why? Thus, even though there are only three paths emanating from point 1, it is that point that is the best place for the tiger to be in terms of avoiding being cornered by the leopards.

Our next example of a game will involve the factor of chance, in the form of having dice-type pieces. It will help us to understand some things about the possible outcomes of different kinds of dice. So, our next activity will be to discuss some ideas from this topic.

A PRIMER ON PROBABILITY

The next example we will discuss has to do with the third type of game, which depends on probability and chance. To understand the calculations, we will have to learn how to calculate certain probabilities. This will be a glimpse at the rather extensive branch of Western mathematics called probability; however, we will see that cultures that develop games involving probabilities often understand many of these same concepts. For our purposes, we will calculate probabilities within the context of numbers of arrangements of different types of objects that can appear. A basic principle of probability is called the *multiplication rule*, which reads like this:

> *If a selection consists of two parts such that one selection can be made in n ways and the other can be made in m ways, then the total selection can be made in $n \cdot m$ ways.*

For example, if there are three ways to select an entree from a menu and two ways to select a side item, then there are three times two, that is, six ways to make a selection consisting of an entree and side item. This concept is true for more than two ways, in the sense that if there are three selections that can be made in n, m, and k ways, then the total selection can be made in $n \cdot m \cdot k$ ways. If there are more ways, the pattern continues in this way. Another principle we will use comes from the concept of making arrangements of objects, also called *permutations*. If we want to arrange seven balls of differing colors, then for the first ball we have seven choices. Once we choose which ball to put in the arrangement, there are six balls left that we can choose from for the second position in the arrangement. After that, there are five balls left, and so on. Applying the multiplication rule, we have seven selections that can be made in seven ways, then six ways, then five, and on down until the seventh position in which we have only one choice. In our Western notation, this calculation is as follows: The total number of ways to arrange seven balls of differing colors is:

$$7 \cdot 6 \cdot 5 \cdot 4 \cdot 3 \cdot 2 \cdot 1.$$

There is a specific notation for this kind of calculation, called the *factorial*, defined as $n!$, that is,

$$n! = n \cdot (n-1) \cdot (n-2) \cdot \cdots \cdot 3 \cdot 2 \cdot 1.$$

1! is defined to be 1 (and 0! is also defined to be 1). For instance, $4! = 4 \cdot 3 \cdot 2 \cdot 1 = 24$. In the previous example, arranging the seven balls can be made in 7! ways (by the way, this comes out to be 5,040 ways).

Now, if we want to think of some of the objects as being indistinguishable, then some of the arrangements are repetitious. For example, if of the seven balls three are green and we decide to consider those three as indistinguishable, then selecting three green balls for the first three positions is only one arrangement; it does not matter if we rearrange the three green balls themselves because we are considering them as indistinguishable. In effect, we divide out all the repetitions. In mathematical notation, this is expressed in the following way:

$$N = \frac{n!}{(n_1)!(n_2)! \cdots (n_r)!},$$

where we consider r groups of objects to be indistinguishable such that n_1 of the objects are indistinguishable, n_2 of the objects are indistinguishable, up to n_r indistinguishable objects. This may seem like a complicated formula; however, most of the quantities are not so difficult to determine. Now is a good time to look at an example of how these numbers are calculated.

Example 7.4 Consider the number of arrangements of the letters of the word "Mississippi." The total number of permutations is 11! because there are 11 letters in the word. Now suppose we choose to consider letters of the same type to be indistinguishable, such as the four letters "s." We need to determine how the indistinguishable letters can be arranged within themselves. This is what the terms $(n_1)!$, $(n_2)!$, and so on are in the expression above for N. We look more closely at the word "Mississippi." Write it by labeling each letter: $Mi_1s_1s_2i_2s_3s_4i_3p_1p_2i_4$. Then we consider i_1, i_2, i_3, i_4 to all be the same, the four letters s_1, s_2, s_3, s_4 to be the same, and the two letters p_1 and p_2 to be the same. There are 4! ways to arrange the letters i_1, i_2,

i_3, i_4, as four objects. There are also 4! ways to arrange the letters *s*, and 2! ways to arrange the letters *p*. There is only 1! way to arrange "M." Thus, the number of arrangements of the letters of the word "Mississippi" is:

$$N = \frac{11!}{(4)!(4)!(2)!1!}.$$

(N comes out to be 34,650.) Although we included it, the term $1! = 1$ was not really necessary.

Now let us consider n objects in which there are two possible outcomes per object. Think of coins. We are interested in the probability of various combinations of these outcomes from rearranging the n objects, which would represent the outcomes from tossing the coins. Using the multiplication rule, there are 2^n possible outcomes, because two possible outcomes for object 1, two for object 2, and so on, until two outcomes for object n, and the ensuing product is: $2 \cdot 2 \cdot 2 \cdot \cdots \cdot 2$ (n times) for a total of 2^n. As an example, if we toss three coins, then there are $2 \cdot 2 \cdot 2 = 2^3 = 8$ possible outcomes. If you want to check this, you could list all eight of them separately, going through the possibilities of heads (H) or tails (T): HHH, HHT, HTT, TTT, HTH, and so forth.

Our next step is to determine the probability of a particular outcome. The basic principle of probability at use here is that *a probability consists of the number of ways a particular outcome can occur divided by the total number of possible outcomes*. We can write probability P of an outcome in generic equation form like this:

$$P = \frac{\text{the number of ways an arrangement is possible}}{\text{the total number of possible outcomes}}.$$

We can connect this with the number N that we looked at earlier, namely,

$$P = \frac{N}{\text{the total number of possible outcomes}}.$$

Because the type of game we will look at below has to do with two-sided figures (with distinct patterns on each side), the analogy of tossing coins is useful again.

Example 7.5 What is the probability of getting three tails after tossing five coins?

To solve this, we use the formula above for the probability P. For each coin there are two possible outcomes and there are five coins, so the total number of outcomes is $2^5 = 32$. For the value of N, note that there are 5! ways to arrange the coins. We consider the outcomes of tails as being indistinguishable; that is, we do not distinguish between the specific three coins that would appear as tails. Thus, there are 3! arrangements of tails. This leaves the remaining two coins as heads and there are 2! ways to arrange those (as indistinguishable). Therefore, we get

$$N = \frac{5!}{3!2!} = 10, \text{ and } P = \frac{10}{2^5} = \frac{10}{32}.$$

We leave the last fraction unreduced in order to compare it with probabilities of other outcomes. Let us look at another example of coin tossing, but this time we will have different types of coins.

Example 7.6 Suppose we have six coins, say, three pennies and three nickels. On a coin toss of all six, what is the probability of having two of the pennies showing tails and one of the nickels as heads?

For this problem, we can consider the outcome of the pennies and the outcome of the nickels as being independent. In other words, we are not including a condition that the outcome of the nickels is in some way dependent on the outcome of the pennies, nor vice-versa. Such a condition would give us a different probability question whose solution would also differ from what we will now calculate. At any rate, the assumption of independence in this problem means we can use the multiplication rule and multiply the probability of the pennies with the probability of the outcome of the nickels. First, the pennies: There are 2! ways to have two tails come up, and then the third penny, assumed to come up heads, occurs in 1! ways. There are a total of 2^3 possible outcomes for the three pennies. This gives us:

$$N_{pennies} = \frac{3!}{2! \cdot 1!} = \frac{3 \cdot 2 \cdot 1}{2} = 3, \text{ and } P_{pennies} = \frac{N_{pennies}}{2^3} = \frac{3}{2^3}.$$

Now we consider the three nickels. Notice that if one of the nickels is heads, then the other two must be tails, so

$$N_{nickels} = \frac{3!}{1! \cdot 2!} = \frac{3 \cdot 2 \cdot 1}{2} = 3 \text{ and } P_{nickels} = \frac{N_{nickels}}{2^3} = \frac{3}{2^3}.$$

The probability of both events occurring is their product:

$$P = P_{pennies} \cdot P_{nickels} = \frac{3}{2^3} \cdot \frac{3}{2^3} = \frac{9}{2^6} = \frac{9}{64}.$$

With these examples, we have enough tools to calculate the probabilities that occur in the next example of a game.

A NATIVE AMERICAN DICE GAME

We will discuss a game called Pay-Gay-Say. It is a game from the Ojibwe culture of North America. It is a game of chance, usually played within the context of gambling. The etymology of "pay-gay-say" comes from the Ojibwe word *pagessewin*, which translates as "dish game." This game has been popular in many Native American cultures, and there are many versions of this game among Native American cultures of North America. Cullen (1975) is a fundamental reference on Native American games, in which you can find a lot of the versions of the dish game that we will now investigate. The description of the game is as follows: Two people play. There is a shallow bowl made from the root of a maple tree. There are eight "dice" pieces, traditionally made from the antlers or bone material of a deer. Each of these eight pieces is plain on one side, and has criss-crossed grooves on the other side. The grooved side is stained black, and the other side remains a light color. The pieces have several shapes. One is shaped like the figure of a person, four are shaped like circles, two are shaped like knives, and one has a shape that resembles a rifle. There are eighty counting sticks that mark points. The sticks may be divided into two sets of forty for each player, or all placed in a common pile.

The game is played as follows. First, there is a prayer to start the game. The eight game pieces are then put in the bowl. Each player takes turns tossing the dice pieces by tipping and releasing the bowl in such a way as to have the pieces go high enough to have the sides be mixed, yet have all pieces fall back into the bowl. Each outcome counts for some number of points that I will describe shortly. The number of points from a toss decides the numbers of sticks that the player takes from either the opponent's pile or the common pile. The person who ends up with all eighty counting sticks wins the game.

In the following description, we will calculate the probabilities of the given outcome and compare it with the number of points that the outcome is worth. The idea is to see if the number of points awarded is comparable to the probability of the given outcome. That is, an outcome that is highly unlikely would count for a lot of points, whereas an outcome that is more likely would count for fewer points. Table 7.2 is a partial list of points awarded for various outcomes. I will use "up" to denote the plain (lighter) side shown, and "down" for the etched side shown.

Let us calculate the probabilities of some of the outcomes in this game. More calculations of probabilities are left to the exercises.

Example 7.7 Calculate the probability of "1 circle up, all other pieces down."

To solve this, we consider two probabilities: One is the probability that one of the four circles is up and the other three down. The second is that all of the

TABLE 7.2 Chart of Points Awarded for Various Outcomes in Pay-Gay-Say

Outcome	Points
1 circle up, all other pieces down	1
1 circle down, all other pieces up	1
2 circles down, all other pieces up	2
3 circles down, all other pieces up	3
4 circles down, all other pieces up	18
4 circles up, all other pieces down	18
All pieces down	18
All pieces up	18
Person, rifle, and one knife up, all others down	9
Person, rifle, and one knife down, all others up	9
Both knives up, all others down	20
Both knives down, all others up	15
Person down, all others up	20
Person up, all others down	40

other four (noncircle) pieces are down. Because these events are independent of each other but must both occur, we multiply their probabilities as we did for the previous pennies and nickels example.

For the one of four circles, we have

$$N = \frac{4!}{1!3!} = 4,$$

and probability

$$P_1 = \frac{4}{2^4} = \frac{4}{16}.$$

For the other four pieces down, we have

$$N = \frac{4!}{4!} = 1 \text{ and } P_2 = \frac{1}{2^4} = \frac{1}{16}.$$

Thus, the total probability is

$$P = \frac{4}{16} \cdot \frac{1}{16} = \frac{4}{256}.$$

Example 7.8 Calculate the probability of "person, rifle, and one knife up, all other pieces down."

To solve this, we consider four probabilities: The probability that the person shows face up is 1/2, the same as flipping a coin. Similarly, for the rifle to show up is 1/2. For one knife to show face up, we could list all possibilities (four total, two of which have one knife up, one down), or we can use the formulas:

$$N_{knife} = \frac{2!}{1!1!} = 2, \text{ and } P_{knife} = \frac{2}{2^2}.$$

For the rest of the pieces, we can consider them as indistinguishable in this context (they are four circles, anyway, in this example), so the probability that all show face down is:

$$N_{circles} = \frac{4!}{4!} = 1, P_{circles} = \frac{1}{2^4}.$$

Thus, the overall probability is

$$P = \frac{1}{2} \cdot \frac{1}{2} \cdot \frac{2}{2^2} \cdot \frac{1}{2^4} = \frac{2}{256}.$$

Notice that we always have $2^8 = 256$ possible outcomes in each overall probability.

Example 7.9 Calculate the probability of the outcome of "four circles down, all others up."

To solve this, we consider two probabilities: One is the probability that all four circles are up. The second is that all of the other four pieces are down. For all four circles up, we have

$$N_{circles} = \frac{4!}{4!} = 1,$$

and probability

$$P_{circles} = \frac{1}{2^4} = \frac{1}{16}.$$

For the other four pieces down, we have

$$N_{others} = \frac{4!}{4!} = 1 \text{ and } P_{other} = \frac{1}{2^4} = \frac{1}{16}.$$

The overall probability is $P = \frac{1}{16} \cdot \frac{1}{16} = \frac{1}{256}$.

THE CULTURAL-MATHEMATICAL ASPECTS OF PAY-GAY-SAY

Now we can look at what the probabilities have to do with the number of points awarded. The outcome of one circle up, all others down, occurs with probability 4/256 and is worth one point. The outcome of the person image, rifle, and one knife up, with all others down, occurs with probability 2/256 and is worth 9 points. Finally, the outcome of four circles down, all others up, occurs with probability 1/256 and is worth 18 points. From these (and checking probabilities of other possible outcomes), we can see a general pattern in which outcomes that are less likely to occur are worth more points and those that are more likely to occur are worth fewer points. In this case, the probabilities, from most to least likely, are 4/256, 2/256, and 1/256, and the number of points awarded to the respective outcomes is 1, 9, and 18. This shows us that as the Ojibwe developed the way the game is played, they were using an understanding of basic ideas of probabilities relative to this kind of game, either from numerous observations of outcomes, or by specific calculations.

As for the cultural aspects of the game, it is known that games such as Pay-Gay-Say have often been played as part of larger social contexts. For example, dish games have existed in many Native American cultures for a long time and are connected with certain rituals, storytelling, and other activities considered important in the particular culture. Often a dish game is played with someone who is gravely ill, not for enjoyment, but as part of a solemn social setting. So, the game of dish is not just a pastime activity. Pasquaretta (2003: 120–125) tells it this way: "Associated with rituals of play and storytelling, games of chance connect the people to their communal origins and destiny. This tradition is most profoundly evident among Haudenosaunee who consider *Gus-ka-eh*, the Sacred Bowl Game, a divine amusement made by the Creator for the happiness of the people."

EXERCISES

Short Answer

7.1 For each of the following combinations, describe an example of a known game or part of a game that satisfies that combination. Explain how the types are part of the scoring or determining the outcome of the game. Do not include actions that determine the start of the game (such as a coin flip); these do not imply that chance is a main component of the game.

(a) A combination of physical skill and strategy.

(b) A combination of strategy and chance.

(c) A combination of physical skill and chance.

(d) Can you find a game that represents a combination of all three: physical skill, strategy, and chance?

7.2 For the Kpelle game, is there another solution to the problem? If so, explain it. If not, explain why not.

7.3 **(a)** Do you know of a version of the Kpelle riddle game? If so, describe how it is similar to the Kpelle game, and also describe the differences between the game you know and the Kpelle game.

(b) Describe the cultural aspects of the game you know and compare it with the Kpelle game.

7.4 For each of the games we have looked at in this chapter (the Kpelle game, Hat Diviyan Keliya, Pay-Gay-Say), explain to which of the broad categories (physical skill, mental skill/strategy, chance, or combinations of these) the game belongs.

7.5 Think back to the motivational question regarding mathematical properties of geometric figures that have certain numbers of paths emanating from the intersections and/or vertices of the figure. Explain briefly the connection between these ideas and ideas of strategy in board games.

Calculations

7.6 **(a)** Suppose that for the Kpelle game we assume that the man can transport two objects at the same time while crossing the river. Make a chart representing a solution to the problem of transporting all three objects to the other side of the river, similar to that shown in Table 7.1. Be sure to explain the importance, if any, of having some steps in which only one object travels with the person.

(b) Suppose in the Kpelle game the man has two leopards that might fight with each other if they are together; otherwise the assumptions of the game are as in Example 7.1. Thus, there are now four objects to be moved instead of three. Determine if there is a solution to the Kpelle game if the man can transport only one object across the river at a time. Then determine a solution to the problem if the man can transport two objects across the river at a time.

7.7 From the Hat Diviyan Keliya game.

(a) Is it possible for the leopards to win the game if there are four remaining? Explain, by indicating in which of the ten points of the game board the tiger and leopards would be in order for this to be either possible or impossible.

(b) From part (a), determine the minimum number of leopards that would have to remain in order for the player managing the leopards to still be in the game. Do this by showing on the game board that if there are fewer leopards, then the tiger cannot be cornered regardless of which point on the board the tiger moves to.

7.8 From the Hat Diviyan Keliya game.

(a) Refer to Example 7.3. Explain the details of that example.

(b) Which point (or points) on the game board are ones that the player with the tiger would want to avoid? Explain your response by showing on the game board how the tiger would be at a disadvantage at this point (or these points) compared with other points on the game board.

7.9 Suppose you have seven pennies for tossing. Determine the value of N and the probabilities of the following outcomes. Explain your steps.

(a) Two tails. **(b)** All heads.

7.10 Suppose you have five pennies, three nickels, and two dimes, and you plan to toss them. Determine the value of N and the probabilities of the following outcomes. Explain your steps.

(a) Two of the pennies come up tails, two of the nickels come up heads, and one dime comes up heads.

(b) One penny comes up heads, and all other coins come up tails.

(c) All coins come up heads.

7.11 Consider a dish-like game having five circular pieces, six triangular pieces and four hexagonal pieces. Determine N and P for the probability of three triangular pieces up, and all the other pieces down.

7.12 From the Pay-Gay-Say game.

(a) Consider the outcome of one knife showing up, all other pieces down. Calculate N and the probability that this outcome would occur.

(b) This outcome gets a value of 12 points. Explain how this point value shows an understanding by the Ojibwe of the probabilities of various outcomes in the game.

7.13 From the Pay-Gay-Say game.

(a) Consider the outcome of the person image showing up, all other pieces down. Calculate N and the probability that this outcome would occur.

(b) Consider the outcome of the person image showing down, all other pieces up. What is the probability of this outcome? Why?

(c) The outcome of part (a) is worth 40 points and the outcome of part (b) is worth 20 points. What does this say about the game of Pay-Gay-Say?

7.14 The Cayuga culture of what is now the northeastern United States, plays a version of the game of dish as follows (see Ascher, 1991: chapter 4): There are six indistinguishable round pieces, made of peach stones that are left plain on one side and burned to be black on the other side. The game is played in a similar fashion (though not identically) as the game of Pay-Gay-Say. Here are a couple details of the game that you can think about.

(a) Consider the outcome of all pieces down (black side up). Calculate N and the probability that this outcome would occur.

(b) Consider the outcome of three pieces up and three pieces down. Calculate N and the probability that this outcome would occur.

(c) The outcome of all pieces down is worth 5 points. The outcome of three pieces up and three down gets no points. Explain how these point values show an understanding by the Cayuga of the probabilities of various outcomes in the game. Compare this with the outcomes and points in Pay-Gay-Say.

Essay and Discussion

7.15 Write about or discuss general ideas about games. Describe the four basic types as discussed in this chapter. Describe at least two examples of games that are of only one type, and at least one example of a game that is a combination of types. Your examples should be different from those discussed here. Explain how the mathematics of a game can imply some mathematical knowledge of the culture that played the game. Use examples distinct from the exercises and examples from this chapter.

7.16 Write about or discuss a Native American version of dish other than the game of Pay-Gay-Say. An excellent source is Cullen (1975). There is also information about the Cayuga dish game in Ascher (1991). Describe the cultural context of the game, including such aspects as whether the game is played under special cultural or social circumstances. Describe how the game is played, how points are awarded, and the probabilities of several specific outcomes. Compare the probabilities with the number of points awarded and make observations about what that information implies about knowledge in the particular culture about probabilities.

7.17 Write about or discuss a game from a part of the world that is far from where you live (e.g., from another continent). Describe the cultural and mathematical aspects of the game as we have done in this chapter. *Note*: This question could be expanded to be a longer essay or term paper project. Also keep in mind that some games include complex mathematics.

7.18 Write about or discuss two or more other tiger–leopard type games. Examples would include Len Choa from Thailand, Pulijudam from India, Rimau-rimau from Malaysia, and many others. Compare aspects such as the geometric shapes of the game boards, the number of vertices and intersections of each type, the number of pieces of each type (tiger-type pieces versus leopard-type pieces), and some similarities and/or differences between the basic rules of each game. Good references for finding information about these games would include Murray (1952: 106–112), Parlett (1999: 193–196), and Zaslavsky (2003: chapter 2).

7.19 Write about or discuss board games around the world other than Hat Diviyan Keliya. Examples could include Mancala-type games, fox and geese games, and so on. The books listed in Exercise 7.17 above are also good sources of information on these types of games. Explain the cultural aspects and considerations of the people who play the game (or games) that you investigate, the mathematical aspects involved, and the connection between these two aspects.

7.20 Write about or discuss comparisons and contrasts between the cultural-mathematical connections of games and the cultural-mathematical connections of divination. Include comments about why certain mathematical concepts are part of both games, and the cultural implications of such similarities.

Additional Comments

- The probability concepts from this chapter represent the tip of an iceberg of mathematical information. Persons with some background in probability and statistics can investigate other games that involve, for example, both strategy and chance, in which probability considerations get more complicated.
- My description of the four broad categories of games is my own invention. There is an entire branch of Western mathematics devoted to game theory, with applications to contexts such as economics. My intention here is to keep the focus on games that are from a cultural context.
- Cullen (1975) contains a huge amount of information about Native American games. Although he does include some culturally relevant comments about the particular Native American culture that played a particular game, I would suggest that if you are going to look into the cultural background of one or more of the games, you should also seek

a general reference on the culture you wish to learn about. Similar comments about the Asian board games discussed here can be made regarding investigations into the cultural aspects of the games.

- The two books on multicultural games by the late Claudia Zaslavsky (Zaslavsky, 1998, 2003) are written to encourage children to learn about mathematics through multicultural games, but as you can see from this chapter, the games can be used as examples of cultural mathematics, and ideas about strategy, teaching of mathematical concepts, and so on, can be built around the context of these games. Plus, they are fun!
- My information on probability comes from an old book by John Freund. A newer version of the book is listed as Freund (1993) in the Bibliography.

8

CALENDARS

Motivational Questions: Before you begin reading this chapter, take some time to think about your responses to the following questions.

- *What is a calendar?*
- *What is a calendar used for?*
- *What is the difference between a lunar cycle and a month?*
- *What kind of information is printed on your (or someone else's) calendar? What kind of information is written on the calendar in handwriting?*
- *What is a leap year? How is it determined? Is there any other way to determine it?*
- *Does the first day of a new year in a calendar have to be the same day? What would happen if each year started on a different day?*

INTRODUCTION

This chapter is about calendars. We will look at the cultural context of calendars and some of the mathematics involved in calendars. We will examine two examples of calendars, that of the *Nuu-chah-nulth* (also called the *Nootka*), and that of the *Mayans*.

Introduction to Cultural Mathematics: With Case Studies in the Otomies and Incas, First Edition.
Thomas E. Gilsdorf.

It probably goes without saying that a calendar is a device for keeping track of time. Although we may think of calendars as being used to measure astronomical events like lunar cycles, in fact the social and cultural aspects of a calendar are at least as important.

If there is a modern calendar you can look at from your home or class (preferably from a previous year), take a look at it. It may show some information about phases of the moon, for example, but most of what is printed on it has to do with social and/or religious events. They could be special holidays like Independence Day, Veterans Day, or Christmas. If the calendar is for a particular place, such as a school, then there are probably other items printed on it, such as teacher development days or conference days. Meanwhile, there are probably items that are written by hand on the calendar, such as birthdays of family members or various kinds of appointments. In effect, most of the information on the calendar is social or cultural.

Our next goal is to discuss the mathematical aspects of calendars. First, we must think about their basic structure. Because of rotations and revolutions—that is, the Earth rotating about its axis, the moon revolving around the Earth, and the Earth and other planets revolving around the sun—the basic concept of a calendar is that of *cycles*. By cycle, I mean some event that repeats. It could be as short as a day, or as long as the time it takes for certain planets to align (which may repeat only after hundreds of years!). There are three types of cycles, and we will now discuss them.

PHYSICAL, NATURAL, AND CULTURAL CYCLES

Let us start with physical and natural cycles. The most common *physical cycles* that appear in calendars are those of the Earth as it rotates about its axis (days), the cycle of the moon as it revolves around the Earth, and the cycles of the Earth and other planets as they revolve around the sun. These physical cycles trigger other *natural cycles* on our planet. Such natural cycles include, for example, migrations of animals, changes in plants, or changes in climate patterns. At any rate, for both physical and natural cycles, humans have no control over them.

On the other hand, there are cycles that are completely determined by people. I will call them *cultural* cycles. An example of a cultural cycle is a week, that is, seven days. What makes this a cultural cycle is the fact that it was determined without dependence on any particular physical or natural cycle. Indeed, at some point our calendar could be changed to have weeks consisting of 8, 10, or 50 days. It all depends on the culture that defines this kind of cycle. Another cultural cycle is the month.

As we think more about calendars, we will see that there are also cycles that are combinations of physical or natural cycles, and cultural cycles. Here is one potential example of such a cycle. Suppose that, instead of declaring the first day of winter to be on the day of the winter solstice, we declared the start of winter to be the first day it snows in a particular place, say, in St. Paul, Minnesota, with the possibility that certain "calendar officials" could adjust that day if necessary. In some years, it could happen that the first snow does not come until very late in the year. This would be a case in which the first day of winter could be changed by the "calendar officials" to an earlier day. In this way, the cycle sometimes depends on a natural cycle—a snowfall—and sometimes is arbitrarily decided by people. Moreover, we will see that *a year in a calendar does not have to start on the same day*. This is something to keep in mind for the rest of this chapter because it turns out that many cultures start a new year using a physical, natural, or cultural cycle, without requiring that the start of the year be on the same day of the year each time. In such cultures, some years can and do have more days to them than other years. Now it is time to look at some examples of calendars. You will notice that the cultural aspects of calendars are part of the discussion right from the beginning.

Before we start our investigation into the mathematics of calendars, it is worthwhile to point out that calendars vary in many creative ways from culture to culture. In fact, there are few rules that have to be followed in constructing a calendar. As we saw in the previous paragraph, the beginning of a new year does not have to be on the same day as the previous years, and there can be cultural factors that dictate arbitrary changes in a calendar. The main importance of a particular calendar is how it is used by the culture that created it.

THE NUU-CHAH-NULTH CALENDAR

Our first example of a calendar comes from the Nuu-chah-nulth (or Nootka) culture of Vancouver Island on the northern Pacific coast of North America, in the Canadian province of British Columbia. Most of my description of the Nuu-chah-nulth calendar comes from Folan (1986), where the culture is referred to as the Nootka. The Nuu-chah-nulth culture has inhabited various parts of the west coast of the island for more than 4,000 years. Their language group is called Wakashan. Until the settling of Europeans on Vancouver Island, the Nuu-chah-nulth were a hunting, fishing and gathering culture. The discussion in Folan (1986) includes descriptions of several interpretations of the Nuu-chah-nulth calendar that have been made by people of varying backgrounds. I will describe the interpretation made by Moziño in 1913, of the Yuquot area of the island. Moziño described the Nuu-chah-nulth calendar as consisting of fourteen cycles of twenty days, such that days could be added to the cycles. How many days would be added to a cycle was decided by the chiefs. Notice that this means each cycle and each year could have a different number of days in it and that the same cycle could have a different number of days in it from one year to the next. Table 8.1 shows a list of the fourteen cycle names, along with the descriptions given by Moziño. For some cycle names, a description was not given, and I have included question marks indicating possible

TABLE 8.1 Nuu-chah-nulth (Nootka) Cycle Names

Cycle Name	Translation/Meaning
Sta-tzimitl	*Wasp (?)*
Tza-quetl-chigl	(Not described)
Ynic-coat-tzimitl	*Dog salmon*
Euz-tzutz	*Rough sea*
Ma-mec-tzu	*Elder sibling*
Cax-la-tic	*Younger sibling*
A-ju-mitl	*Great cold (?)*
Vat-tzo	(Not described)
Vya-ca-milks	*Herring fishing*
Aya-ca-milks	*Herring spawning*
Ou-cu-migl	*Geese*
Ca-yu-milks	*Religious festivals and whaling*
Ca-huetz-mitl	*Salmonberry*
Atzetz-tzimil	*Fruit, roots, shoots, leaves, and flower collecting*

alternative descriptions, as described by Folan. The first cycle of the Yuquot (Nootka) year begins at the time that corresponds roughly to July.

Let us look deeper into the description of the Nuu-chah-nulth calendar. First, we can see that several of the cycles are connected with natural cycles. Nevertheless, the fact that the chiefs would frequently add days to the cycle means there is a cultural aspect of the cycles as well. To the first cycle, *Sta-tzimitl*, many days were typically added to its original twenty days, depending on factors such as the availability of fish (halibut, tuna, cod, and bream) that could be caught then (Folan, 1986: 96). So, this cycle represents an example of a natural cycle (movement of fish) mixed with a cultural cycle (arbitrarily changing the length of the cycle).

Now, this calendar may seem quite different from the kind of calendar we are used to using. Notice that there is much less emphasis on details of specific days and on keeping cycles the same length. On the other hand, the calendar was useful to the Nuu-chah-nulth culture, and this is the important point. It is a flexible calendar that adjusts easily to minor fluctuations in natural cycles. Also, observe how this example shows us the difference in perceptions of time. Recall our discussion from Chapter 1 about the Western emphasis on number. In this context, that emphasis on number manifests itself as a difference in emphasis of time between the Nuu-chah-nulth culture and Western culture.

MATHEMATICS OF CYCLES

To prepare for looking at a calendar that includes numerous cycle calculations, here are some facts that are often useful in discussions about calendars. One *lunar cycle* is a complete cycle of the moon through its phases, that is, from full moon to full moon (although it could be measured starting from any part of this physical cycle). Many cultures have had accurate knowledge of lunar cycles. Our modern data on lunar cycles is that one lunar cycle lasts approximately 29.531 days. The next cycle is that of the Earth revolving around the sun. I will refer to it as the *solar year*, although it is also called the *tropical* year. This physical cycle lasts approximately 365.2422 days, which is approximately 12.368 lunar cycles.

Next, a cultural fact is to be noted. In Western culture we are used to the idea that each calendar year starts at the same time during the solar year. In other words, the assumption is that each year starts on the same day of the solar year cycle. In fact, we adjust our calendar from time to time to make sure it coincides with the solar year cycle.

The first mathematical structure we will look at is one that measures when cycles coincide on the same day. I will follow Ascher (2002) and will call these *supracycles*. If a calendar has one cycle of eight days and another cycle of eleven days, then if the two cycles coincide today, it will take 88 days for the cycles to coincide again. This is because 88 represents eleven complete cycles of eight days and eight complete cycles of eleven days. In Western mathematical notation, what we just calculated is called the *least common multiple* of the two numbers representing the cycles. The least common multiple of two or more integers is abbreviated by *lcm*. Hence, the least common multiple of two numbers m and n would be denoted by $lcm(m, n)$. For more than two numbers, say, m_1, m_2, up to m_j, we would denote their least common multiple by $lcm(m_1, m_2, \ldots, m_j)$. There is another issue regarding coincidences of cycles. If we assume the cycles are independent of each other, then the cycles will coincide after the number of days representing their *lcm* has elapsed. On the other hand, if one cycle always finishes before the other cycle advances, then the number of days until they coincide will be the product of the numbers, not their *lcm*. Here is an example.

Example 8.1 Determine when two cycles, one of twelve days, another of fifty days, would coincide if:

(a) The two cycles are independent.

(b) The twelve-day cycle makes a complete run before the fifty-day cycle advances.

(c) The fifty-day cycle makes a complete run before the twelve-day cycle advances.

To solve part (a), we calculate $lcm(12,50)$. You probably recall that the process is to write each number as product of its prime factors. It goes like this:

$$lcm(12, 50) = lcm(2^2 3, 5^2 2) = 2^2 \cdot 3 \cdot 5^2 = 300 \text{ days.}$$

For part (b), we calculate the product: $12 \cdot 50 = 600$ days. Part (c), although appearing to be different from part (b), is the same calculation, and we obtain 600 days again.

Another mathematical concept that appears in many calculations relevant to calendars is what in Western mathematics is called *congruence modulo n*, part of the topic of *modular arithmetic*. If you have read Chapter 6, you will recall the discussion of calculations modulo 4 from the divination process from the Caroline Islands. Whether you have read Chapter 6 or not, I will define the terms again below for convenience. The following motivates us to see how this kind of calculation gets used.

Example 8.2 Suppose January 1 is on a Thursday. Which other days in January will also be on Thursdays?

Of course, we can look at a calendar and see that it would be on January 8, January 15, and so on. There is another way to look at it. From January 1 to January 8 is one week, a cycle of seven days. To January 15 is two cycles of seven days, January 22 is three cycles, and so forth. If we write this information out as equations, it looks like this: $8 - 1 = 7, 15 - 1 = 14, 22 - 1 = 21$. In general, day a will be on the same day as January 1 whenever the difference between a and 1 is a multiple of seven. In other words, $a - 1 = k \cdot 7$ for some integer k. We do not really need this level of sophistication for determining when certain dates are on the same day, but we will see that in calendars with cycles of large numbers, this kind of calculation becomes a useful tool. In Western mathematics, this calculation is referred to as calculating *congruences modulo n*. We will calculate many quantities this way, but first I would like to write out a common form of the (Western) definition and notation:

Definition 8.1 Suppose n is a fixed integer greater than one. Two integers a and b are called *congruent modulo n* if their difference is a multiple of n. It is written like this: $a \equiv b \pmod{n}$ if $a - b = k \cdot n$ for some integer k.

An explanation is in order. First, all quantities are integers in the definition. Second, the values, a, b, and k can be negative. We will see situations of negatives with calendars later.

In the previous comments about days of the week in January, we can now view them as describing this: Day a is on the same day of the week (within the same month) as day b if $a \equiv b \pmod{7}$. Indeed, $22 - 1 = 21 = 7 \times 3$ is equivalent to the statement that $22 \equiv 1 \pmod{7}$. More examples are next.

Example 8.3 Suppose a calendar has cycles of twenty days.

(a) If today is day three, will day 158 be on the same cycle-day as today?

(b) Determine two distinct (positive) days that are on the same cycle-day as day thirteen.

Notice that instead of giving the twenty-day cycle a name such as "month" or "week," we can call it a cycle. Now, for part (a), let us take the difference: $158 - 3 = 155$. The number 155 is clearly not divisible by 20, so the answer to part (a) is "No, day 158 is not on the same cycle-day as day three." For part (b), we need to solve the equation $x \equiv 13 \pmod{20}$ for x. The definition of congruence modulo 20 tells us we need to solve $x - 13 = k \cdot 20$ for x. Which value of k do we choose? We need two distinct days, so we just need to choose two distinct values of k. For $k = 1$, we solve $x - 13 = 20$, giving $x = 33$. For the second day, we could choose $k = 2$; however, any other positive integer k will work. For instance, the choice of $k = 5$ gives us $x - 13 = 100$, for which $x = 113$.

The process of determining a value x that satisfies a particular relation modulo n occurs often enough that it is worthwhile to study it a bit more.

Example 8.4 Determine the smallest positive (integer) value of x such that $1002 \equiv x \pmod{365}$.

Symbolically, the definition tells us that this is an equation that looks like: $1002 - x = k \cdot 365$, and we want to determine a value of k that will give us the desired value of x. We need to know how many repetitions of 365 are in the number 1002, and then determine what is left over (the remainder). Although we could simply divide 1002 by 365 with a calculator, cultures that made this kind of calculation did not calculate that way. We can reason our way through the problem as follows: Without a calculator, $365 \times 2 = 730$, and $365 \times 3 = 1{,}095$. This means 1002 is more than twice 365 but is less than three times 365. In Western notation, this amounts to rewriting $1002 - x = k \cdot 365$ as $x = 1002 - k \cdot 365$, by moving around terms. If we try $k = 3$, we have $1002 - 3 \cdot 365 = 1002 - 1095 < 0$. Because we want x to be positive, we choose the next smaller k, that is, $k = 2$. Then we get $x = 1002 - 730 = 272$. We can check this solution to see if it works: $1002 - 272 = 730 = 2 \cdot 365$. This last statement is an expression of the fact that the difference between 1002 and 272 is a multiple of 365, that is, $1002 \equiv 272 \pmod{365}$ and that $x = 272$ represents the smallest positive value that satisfies the modular equation we sought to solve. The calendar interpretation of this example is that we can say: 1002 days from today (assuming no leap year is involved) is two years and 272 days.

Another remark is in order regarding the choices of values. Consider that $52 = 4 \times 13$ within the context of values congruent modulo 13. We know 52 is already a multiple of 13, so there are two natural ways to express this with the modular notation. One is to say that $52 \equiv 13 \pmod{13}$, the other is $52 \equiv 0 \pmod{13}$. In general, for the value n, congruence of a number modulo n can be expressed as an integer value that ranges from 0 to $n - 1$, or as an integer value that ranges from 1 to n. For our discussion of calendars, I am going to use the latter expression. So, we would write $52 \equiv 13 \pmod{13}$ in this setting.

Here is a property of congruence that might be useful to us from time to time:

$$\text{If } a \equiv b \pmod{n}, \text{ then } b \equiv a \pmod{n}.$$

To see how this can be, suppose that $a \equiv b \pmod{n}$. Then $a - b = k \cdot n$, for some integer k. Multiplying both sides by -1, $-(a - b) = -k \cdot n = (-k)\, n$, which means $-a + b = (-k) \cdot n$, or, $b - a = (-k) \cdot n$. Maybe it does not seem right to have $-k$ in the expression, but recall that k can be either positive or negative, so $-k$ is a legitimate choice here. Thus, we can say that $b \equiv a \pmod{n}$. In the exercises, you can learn about more properties of congruence modulo n. In the example above, we could have stated the congruence as $x \equiv 1002 \pmod{365}$, instead of $1002 \equiv x \pmod{365}$.

Next in our list of congruence examples has to do with negative numbers. It corresponds to when we want to calculate a date from the past. Although we could use negatives in our final expression, it is not natural. We do not talk about "May –19," for example. What we have to do is calculate in such a way as to have the final value be positive. Take a look at the next example.

Example 8.5 Determine the smallest positive value of x such that $-29 \equiv x \pmod{12}$.

The problem amounts to making the statement

$$-29 - x = k \cdot 12,$$

where we seek the smallest *positive* integer x that makes the equation true for an integer k. If we move -29 to the other side of the equation, we get: $-x = 12 \cdot k + 29$, and then, $x = 12 \cdot (-k) - 29$. Now, the original k is negative (can you explain why?), so $-k$ must be positive. Let us relabel that "$-k$" as m. Our equation now reads: Determine the *smallest* (integer) value of x such that $x = 12 \cdot m - 29$ is positive. To finish, we multiply 12 by several positive integer values (choices of m) until we find one that gives us the desired x. Just by thinking about multiples of 12 for a moment, if we try $m = 1$, we get $x = 12 \cdot 2 - 29 = -5$. We need a positive result, so we increase m, and realize that $m = 3$ will work, yielding $x = 12 \cdot 3 - 29 = 36 - 29 = 7$. You can check that $7 - (-29) = 36$, which is a multiple of 12. Thus, the solution to $-29 \equiv x \pmod{12}$ is $x = 7$.

It will help us to work through another example, but before that, let us organize our thoughts about congruences with negatives:

- When we say that we want "the smallest value of x such that $x = 12 \cdot m - 29 > 0$," in fact, what we seek is the smallest value of x such that $x = 12 \cdot m - 29$ is between 1 and 12. We could have chosen $m = 4$, to obtain $x = 12 \cdot 4 - 29 = 48 - 29 = 19$; however, the value of 19 is not between 1 and 12, so our choice of x is too big.
- We must be careful with negatives and congruences. It may seem tempting to solve the problem for positive values, but this does not give us the

value we seek. Consider the example of $-7 \equiv x \pmod{5}$. We get the equation $x = 5 \cdot m - 7$, and by choosing $m = 2$, we get $x = 5 \cdot 2 - 7 = 3$, so $x = 3$ is the correct value. On the other hand, we could think optimistically that it suffices to solve the equation $7 \equiv x \pmod{5}$ instead. However, if we change the sign of 7, then 3 is no longer the desired solution, because $7 \equiv 3 \pmod{5}$ would mean that $7 - 3 = 4$ is a multiple of 5; but of course 4 is not a multiple of 5.

- In light of the description above about $-7 \equiv x \pmod{5}$, it would be helpful to have a consistent expression for calculating congruences with negatives. Here is the one I have found to seem most natural: For $x \equiv -A \pmod{n}$ (or equivalently, $-A \equiv x \pmod{n}$), where $-A$ is negative (i.e., A is positive), the expression for x can be obtained by:

$$x = n \cdot m - A,$$

and we seek the (positive) integer m that gives us a value of x that is between 1 and n.

- For larger values it is no longer practical to simply multiply n by integers until we obtain x. Instead, we can divide by n to get a starting point. We will see this soon, in another example.
- *Cultural context*: There are many ways to think about congruences modulo n, involving properties of multiplication and division of numbers. Books on number theory typically discuss some of these aspects. In particular, we can say that the expression $x = n \cdot m - A$ explained above is not the only way to determine the solution x. Within our context of the Mayan calendar, it is well known that the Mayans were experts at solving problems that we are describing with congruences modulo n. Undoubtedly, the Mayans knew many properties of congruences, but they did not express them the way we are expressing them. In some cases, we do not know how they solved the problems; we only know that they did indeed solve them. See Ascher (2002: 62–74) for details of Mayan solutions to some complicated problems related to their calendar system.

Now we are ready for another example.

Example 8.6 Determine the smallest positive value of x such that $x \equiv -100 \pmod{17}$.

Using the format of $x = n \cdot m - A$, we can write this as $x = 17 \cdot m - 100$, and our goal is to determine x that is between 1 and 17. To get a starting value for m, we calculate $100/17 \approx 5.88$, so we can start with $m = 5$. Note that $17 \times 5 = 85$, which is less than 100, so, in fact, we need the *next* value of m, namely, $m = 6$. We get $17 \times 6 = 102$, so, $x = 17 \cdot 6 - 100 = 102 - 100 = 2$, which tells us that $x = 2$.

With the ideas about congruences, we have enough information to study the Mayan calendar system. Let us begin.

THE MAYAN CALENDAR: TZOLKIN AND VAGUE YEAR

The Mayans, who have inhabited an area ranging from southern Mexico to Honduras for over 2,000 years, developed a complex calendar system consisting of many cycles. The calendar has been used for determining cultural events and also for determining when to plant or harvest important crops that have traditionally been part of the cultures of that part of the world (such as corn, beans, squash, herbs and spices, and various grains). We will learn something about a few of cycles of their calendar. The significance of each day in the Mayan calendar system depends on the details of which parts of which cycle it represents. Moreover, the days on which major religious and cultural events occur are determined by careful calculations of cycles. In parts of traditional Mayan territory such as in Guatemala, local events are still determined using a form of the calendar system that we are going to look at. The first two cycles we will consider are described next.

The first calendar cycle, a culturally determined one, is a supracycle of two cycles, of thirteen and twenty days. It is called the *Tzolkin*. Another cycle of 365 days is called a *Vague Year*. We will first look at the Tzolkin cycle, then the Vague Year cycles, and, finally, these two cycles together.

The word "Tzolkin" was denoted by specialists of Mayan culture, and translates as "the counting of days." This supracycle is also referred to as the *Sacred Round*. It is the main calendar cycle for religious and cultural events and, as such, is sometimes also referred to as the *ritual calendar*. The thirteen days are denoted by numbers one to thirteen. The twenty days are denoted by symbols, which are called *glyphs*. Each glyph has a specific name to it, and refers to one of the many gods that are part of Mayan cosmology. Table 8.2 shows the 20 names used in the Tzolkin calendar.

I will use a version of Ascher's notation for the Tzolkin cycle (Ascher, 2002), namely, ${}_iT_j$, where $i = 1, 2, \ldots, 13$, and $j = 1, 2, \ldots, 20$.

TABLE 8.2 Twenty Mayan Day Names for Tzolkin Dates

Number and Name	Number and Name
1 *Imix*	11 *Chuen*
2 *Ik*	12 *Eb*
3 *Akbal*	13 *Ben*
4 *Kan*	14 *Ix*
5 *Chicchan*	15 *Men*
6 *Cimi*	16 *Cib*
7 *Manik*	17 *Caban*
8 *Lamat*	18 *Eznab*
9 *Muluc*	19 *Cauac*
10 *Oc*	20 *Ahau*

We will use numerical symbols for the calculations, keeping in mind that the Mayans did not write the values in this way. For example, the Tzolkin date of 12 *Cib* has the notation ${}_{12}T_{16}$ in our symbolism, and 1 *Kan* is ${}_{1}T_{4}$. Note that the supracycle here is $lcm(13,20) = 260$ days, so a Sacred Round consists of 260 days. It turns out that each cycle, that of thirteen days and that of twenty days, run independently. Here is an example.

Example 8.7 Count the days in the Tzolkin cycle with our notation starting with the day 5 *Lamat*. This day is denoted by ${}_{5}T_{8}$.

We get the following sequence of days: ${}_{5}T_{8}$, ${}_{6}T_{9}$, ${}_{7}T_{10}$, ... When we get to a value of either 13 for i or 20 for j, the counting starts over. Thus, as we continue counting we have the following: ${}_{8}T_{11}$, ${}_{9}T_{12}$, ${}_{10}T_{13}$, ${}_{11}T_{14}$, ${}_{12}T_{15}$, ${}_{13}T_{16}$, ${}_{1}T_{17}$, ${}_{2}T_{18}$, ...

Observe how the i index started over with 1 after it had a value of 13. Continuing, we get: ${}_{2}T_{18}$, ${}_{3}T_{19}$, ${}_{4}T_{20}$, ${}_{5}T_{1}$, ${}_{6}T_{2}$, ...

I hope you can see that the counting is done modulo 13 with i, and modulo 20 with j. Our next goal will be to see how to determine a Tzolkin date that is N days in the future or in the past. Because each cycle runs independently of the other, we can determine the i and j values as two separate calculations. Example 8.8 below gives you an idea of how it works.

Example 8.8 If the Tzolkin date of today is 6 *Oc*, what will the Tzolkin date be that corresponds to 200 days from today?

First, note that in our notation, 6 *Oc* is ${}_{6}T_{10}$. We want to determine the new values, i_N and j_N, where $N = 200$. There are two calculations, one of them being $i_N = (i + N)(\text{mod } 13) = (6 + 200)(\text{mod } 13)$. So, we look at the value of 206. You can check that $13 \times 15 = 195$ and $206 - 195 = 11$, which implies $i_N = 206\ (\text{mod } 13) = 11$. Similarly,

$$j_N = (j+N)(\text{mod } 20) = (10+200)(\text{mod } 20) = (210)(\text{mod } 20).$$

Because 210 is 10 more than 200, we get $j_N = 10$. Hence, the new Tzolkin date is ${}_{11}T_{10}$, which we write as 11 *Oc*.

It will be useful to us to have the congruence (modulo 13 and 20) equations handy, so let us put them down here.

For N days after or before the Tzolkin date of ${}_{i}T_{j}$, calculate:

$$i_N = (i+N)(\text{mod } 13),$$
$$j_N = (j+N)(\text{mod } 20).$$

Let us try another example.

Example 8.9 What was the Tzolkin date 365 days *before* the Tzolkin date of 9 *Akbal*?

As with the previous example, we calculate the new i value modulo 13 and the new j value modulo 20, noting that 9 *Akbal* is ${}_9T_3$. However, this time the value of N is negative because we are looking at a date in the past. We can do the new i value first:

$$i_N = (9-365)(\text{mod } 13) = -356 \ (\text{mod } 13).$$

Using the generic expression

$$x = n \cdot m - A,$$

with $n = 13$ and $A = 356$,

$$x = 13 \cdot m - 356.$$

We are trying to determine x by finding the m value that multiplies with 13 to give a positive integer between 1 and 13. That is, we work with the expression $x = 13 \cdot m - 356$. We want to find the integer x between 1 and 13 that satisfies the equation $x = 13 \cdot m - 356$. We know that $13 \cdot 30 = 390$, but $390 - 356 = 34$, which is not between 1 and 13. If we notice that 34 is a little larger than 26, it tells us that we should try about two multiples of 13 less than 30. So, using 28, we get $13 \cdot 28 = 364$. Then, we calculate: $364 - 356 = 8$. This tells us that the new i value is 8, that is, $i_N = 8$. For the j value, we solve the congruence

$$j_N = (3-365)(\text{mod } 20) = -362 \ (\text{mod } 20).$$

From the expression $j_N = 20 \cdot m - 362$, we observe that what we need for j_N is the multiple of 20 that is just larger than 362. That value is 380, so $j_N = 380 - 362 = 18$. The new j_N is 18, and our final expression for the Tzolkin date is ${}_8T_{18}$, which is 8 *Eznab*. This represents the Tzolkin date that occurred one year before 9 *Akbal*, i.e., one Vague Year before ${}_9T_3$.

Cultural context: You may have noticed that we could have solved the previous problem using a calculator; for example, for $i_N = -365$ (mod 13), we could divide 356 by 13, getting $356/13 \approx 27.38$, which would tell us that if we use the next highest integer greater than 27.38, it will give us a value for x. In fact, that next integer is 28, and we get $13 \cdot 28 = 364$ as before. You may wish to use this technique to check your calculations, but I did not use it here, because this is definitely not how the Mayans would have solved the problem of calculating Tzolkin dates. We do not know how the Mayans solved these problems, only that they did indeed solve such problems.

Let us calculate another Tzolkin date from the past.

Example 8.10 If today has Tzolkin date 7 *Ben*, that is, ${}_7T_{13}$, what was the Tzolkin date two Vague Years ago?

We seek i_N and j_N, for $N = -730$. (Can you explain where the value of −730 comes from?) For i_N, calculate this way:

$$i_N = (7 - 730)(\text{mod } 13) = (-723)(\text{mod } 13).$$

Comparing with the previous example, we solve

$$x = 13 \cdot m - 723$$

by trying to determine the m value that will give us an x value between 1 and 13. Note that $13 \cdot 50 = 650$, and if we add another $13 \cdot 5 = 65$, we get a total of $650 + 65 = 715 = 13 \cdot 55$. This value is still a little less than 723, so we put in one more 13, that is, $13 \cdot 56 = 728$ gives the first integer value larger than 723. In other words, $m = 56$, and then

$$x = 13 \cdot 56 - 723 = 728 - 723 = 5.$$

Thus, $i_N = 5$. You may consider determining x by calculator, in which the calculations go like this: $723/13 \approx 55.6$, which means we use the next integer value larger than 55.6, that is, $m = 56$. This gives us the value of i_N as 5 again. Let me remind you once again that the Maya did not have calculators, so I have presented a non-calculator way to determine i_N in order to keep things a little more realistic.

For j_N, we have

$$j_N = (13 - 730)(\text{mod } 20) = (-717)(\text{mod } 20).$$

Notice that −717 is +3 from −720, which is a multiple of 20. That is, $-717 - (-720) = 3$. Alternatively, we can calculate $j_N = 20k - 717$, and notice that $20 \times 36 = 720$, which gives us $j_N = 720 - 717 = 3$. So, j_N is 3, and the Tzolkin date, two Vague Years ago, is 5 *Akbal*, which represents ${}_5T_3$.

We will see more examples of calculating Tzolkin dates shortly. Because the Tzolkin dates are typically part of an expression that involves the 365-day cycle, it is time for us to look at that cycle. Recall that the 365-day cycle is referred to as the Vague Year. Notice that in Example 8.8, we calculated the Tzolkin date that is one Vague Year before ${}_9T_3$ (which turned out to be ${}_8T_{18}$). The mathematical structure of the Vague Year is found in a calculation of $18 \times 20 + 5 = 365$. There are 18 cycles of 20 days each, then five days are added at the end. To understand how the counting is done, we must take into account the fact that the cycles of 20 occur *within* the cycles of 18. By our earlier discussion of supracycles, we know this means the supracycle for the 18 cycles of 20 is $18 \times 20 = 360$, and not 180 (180 is the least common multiple of 18 and 20). Also, before we start with the notation, we will need to denote the five extra days added at the end. This is done in Western notation by simply marking them as numbers one through five. Finally, as with the Tzolkin

TABLE 8.3 Eighteen Mayan Day Names of the Vague Year

Number and Name	Number and Name
1 *Pop*	10 *Yax*
2 *Uo*	11 *Zac*
3 *Zip*	12 *Ceh*
4 *Zotz*	13 *Mac*
5 *Tzec*	14 *Kankin*
6 *Xul*	15 *Muan*
7 *Yaxkin*	16 *Pax*
8 *Mol*	17 *Kayab*
9 *Chen*	18 *Cumku*

supracycle, the Mayans used numbers for one cycle and glyphs for the other. In the case of the Vague Year, the cycles of 20 are denoted by numbers and the cycles of 18 by glyphs. See Table 8.3 for a list of the 18 glyph names of the Vague Year cycle.

I will use a version of Ascher's notation for the Vague Year (Ascher, 2002) as follows: A day in the Vague Year is denoted as:

$$_{m}V_{l},$$

where $m = 1, 2, 3, \ldots, 20$, and $l = 1, 2, 3, \ldots, 18$.

Here are a couple of examples of how Mayan Vague Year dates look in this notation: We will write 19 *Mol* as $_{19}V_8$, while 2 *Muan* is $_2V_{15}$.

Keeping in mind that the 20-day cycles run *within* the 18-day cycles, the notation for first 360 days of the Vague Year has the structure shown below.

$$_1V_1, {}_2V_1, {}_3V_1, \cdots, {}_{19}V_1, {}_{20}V_1, {}_1V_2, {}_2V_2, {}_3V_2, {}_4V_2, \cdots, {}_{19}V_2, {}_{20}V_2, {}_1V_3, {}_2V_3, {}_3V_3, \cdots,$$

until the last five days, which have the (Western) notation of simply 1, 2, 3, 4, and 5. See also Example 8.11, which follows this discussion.

It turns out that the Mayans denoted every day of a Vague Year by both a Tzolkin date and a Vague Year date. For this notation, I will write

$$({}_iT_j, {}_mV_l),$$

where $i = 1, 2, 3, \ldots, 13$, $j = 1, 2, 3, \ldots, 20$, $m = 1, 2, 3, \ldots, 20$, and $l = 1, 2, 3, \ldots, 18$.

Here are a couple of examples of this notation (see Tables 8.1 and 8.2): (13 *Men*, 9 *Pax*) = $({}_{13}T_{15}, {}_9V_{16})$, and (5 *Chicchen*, 20 *Xul*) = $({}_5T_5, {}_{20}V_6)$.

This way of expressing Mayans dates is called the *Calendar Round* by Mayan specialists. To give you an idea of how this notation works, and so you can see how the notation works for the last five days of a Vague Year, take a look at the following example.

Example 8.11 Here is the notation for the last seven days of a Vague Year.

$$({}_2T_8, {}_{19}V_{18})({}_3T_9, {}_{20}V_{18})({}_4T_{10}, 1)({}_5T_{11}, 2)({}_6T_{12}, 3),$$
$$({}_7T_{13}, 4)({}_8T_{14}, 5)({}_9T_{15}, {}_1V_1)({}_{10}T_{16}, {}_2V_1)\cdots$$

Notice that on the 360th day of a given Vague Year, the Tzolkin date need not be ${}_3T_9$, but the Vague Year date would always be ${}_{20}V_{18}$, just before we count the last five days of a year.

To calculate a number of days in the past or future from a given Vague Year, the calculation is not so simple because of the way one cycle runs within the other, and because of the fact that if the number of days puts us between 361 and 365, we must change the notation. We do not know how the Mayans made the calculations, but in Western notation, it comes out to be as follows (see Ascher, 2002: 69).

Suppose we want to know the Vague Year date that is N days from a given Vague Year date ${}_mV_l$. First, we must calculate a modified value of N and look to see if it is between 361 and 365. We count total days, starting from the beginning of the year, taking into account the possibility that the end value might be more than 365 (in which case we adjust it mod 365). We will give the name N' to the new value. The formula is this:

$$N' = [20(l-1)+m+N](\text{mod } 365),$$

where l and m are the indices of the original date given as ${}_mV_l$.

You can see that the formula counts the number of 20-day cycles given by l (the –1 accounts for the fact that in the first twenty days we have gone through zero cycles of 20), includes the number m of days from the given 20-day cycle, and adds it all to N. To get the new m and l values, denoted by $m'$and l', respectively, we now calculate as follows:

$$m' = N'(\text{mod } 20), \text{and } l' = \frac{N'-m'+20}{20}.$$

Now is a good time to look at an example.

Example 8.12 What is the Vague Year date that is 100 days after 7 *Muan*?

To begin with, we have the notation ${}_7V_{15}$ for 7 *Muan*. In this question, $N = 100$, $l = 15$, $m = 7$, which means

$$N' = [20(15-1)+7+100](\text{mod } 365) = [387](\text{mod } 365) = 22,$$

because 387 – 365 = 22. Then, $m' = 22 \pmod{20} = 2$ and

$$l' = \frac{22-2+20}{20} = 2.$$

The sought-after Vague Year date is 2 *Uo*, considering that this is how the Mayan terms express ${}_2V_2$.

Another example follows. To be consistent with the fact that the Mayans denoted typical days with both a Tzolkin and Vague Year date, we will do the same in the next example (but keep in mind that they did not use the notation we will use).

Example 8.13 What is the Calendar Round date that is 545 days after (12 *Caban*, 4 *Yax*)?

Using Tables 8.2 and 8.3 to write the notation, we get: (12 *Caban*, 4 *Yax*) = (${}_{12}T_{17}$, ${}_4V_{10}$). Let's do the Tzolkin part first. We have $N = 545$, and we then determine i_N and j_N.

$$i_N = (12+545)(\text{mod } 13) = (557)(\text{mod } 13).$$

We can reason that there must be at least forty 13s in 557. Checking some products: 13 × 40 = 520; 13 × 42 = 520 + 26 = 546; 13 × 43 = 546 + 13 = 559. These values tell us that 546 is the closest value to 557 that is still less than 557. Hence, the new i_N is 557 – 42 · 13 = 11. Next,

$$j_N = (17+545)(\text{mod } 20) = (562)(\text{mod } 20).$$

The value 562 is two more than 560, which is a multiple of 20, so $j_N = 2$. At the moment, we now have the new Tzolkin date of ${}_{11}T_2$. We finish by calculating the new Vague Year part, starting with the given values of $N = 545$, $l = 10$, $m = 4$, obtaining

$$N' = [20(10-1)+4+545](\text{mod } 365) = [729](\text{mod } 365) = 364,$$

because 729 is one less than 730, which is twice 365. In other words, if we use 365 × 2 = 730, we have that 730 > 729, so we use 365 × 1 instead. Then $N' = 729 - 1 \cdot 365 = 364$. The value of N' being 364 means the Vague Year part of the the Calendar Round date is the fourth extra day, which we simply write as "4." Our final notation of the new Calendar Round date is:(${}_{11}T_2$, 4), or (11 *Ik*, 4).

We can practice these kinds of calculations via another example.

Example 8.14 Suppose today has a Calendar Round date of (9 *Manik*, 19 *Pax*). Determine the Calendar Round date of 400 days ago.

First, we write the notation for this date, using Tables 8.2 and 8.3: (9 *Manik*, 19 *Pax*) = $({}_9T_7, {}_{19}V_{16})$. We again work on the Tzolkin date first. First, i_N: Using –400 for a day from the past, we have

$$i_N = (9-400)(\text{mod } 13) = (-391)(\text{mod } 13).$$

We use the expression for x that will give us a value that is between 1 and 13:

$$x = 13k - 391.$$

Because $13 \times 30 = 390$, but we seek a multiple of 13 that is larger than 391, we choose k to be the next multiple of 13, namely, 31. In other words,

$$x = 13 \cdot 31 - 391 = 403 - 391 = 12,$$

which gives us the value of $i_N = 12$.

Now for j_N. We solve $j_N = (7 - 400)(\text{mod } 20) = -393$ (mod 20). With our generic formula for x, we solve $x = 20k - 393$. The closest multiple of 20 to 393 is 400, so we calculate $x = 400 - 393 = 7$. Although this may seem surprising, we can observe that because 400 is already a multiple of 20, for any initial value of j (in this case $j = 7$), the new j_N will be the same value again. Thus, for instance, if $j = 11$ and $N = 540$ (or –540), the new j_N will be 11 again because 540 is a multiple of 20.

Next, we work on the Vague Year date. First we determine N'. Our values are $N = -400$, $l = 16$, and $m = 19$. We get N' by the following calculation:

$$\begin{aligned} N' &= [20 \cdot (16-1) + 19 + -400](\text{mod } 365) \\ &= -81\,(\text{mod } 365). \end{aligned}$$

Using our formula for x as before (for negatives with modular arithmetic):

$$x = 365 \cdot k - 81 = 284,$$

where we have used $k = 1$. We obtain m' and then l':

$$m' = 284\,(\text{mod } 20).$$

Recalling how these equations are with positive values, we write the equation:

$$284 = m'(\text{mod } 20),$$

then solving $284 - m' = 20 \cdot k$, which, by inspection, gives us $m' = 4$ because $284 - 4 = 280$ and 280 is a multiple of 20. Another way of obtaining m' is to consider that we want $m' - 284$ to be a multiple of 20. Observing that –284 is only 4 in difference from –280 (that is, 4 – 284 = –280), and that –280 is a multiple of 20 (remember that the multiple can be negative as long as it is an integer), we obtain $m' = 4$. Now for l':

$$l' = \frac{284 - 4 + 20}{20} = \frac{300}{20} = 15.$$

Let us report our final, new Calendar Round date, which represents 400 days before the Calendar Round date of (9 *Manik*, 19 *Pax*). The new Calendar Round date is: (12 *Manik*, 4 *Muan*), which we can write as $({}_{12}T_7, {}_4V_{15})$.

As a summary of how to calculate dates as equations in modular arithmetic, we can state that our Western method of solving $x = A(\text{mod } n)$ is to create an equation based on whether A is positive or negative:

$$x = A - n \cdot k, \text{for } A > 0,$$
$$x = n \cdot k - A, \text{for } A < 0,$$

where we seek integers k whose multiples of n will give us a value of x that is in the range of:

$$1 \leq x \leq n.$$

The last piece of information to look at in this section is the supracycle of Tzolkin and Vague Year cycles. As we know, and the Mayans knew, the length of the supracycle can be obtained by calculating what we refer to as the least common multiple of the two cycles, so it is:

$$lcm(260, 365) = lcm(13 \cdot 20, 5 \cdot 73) = 13 \cdot 5 \cdot 4 \cdot 73 = 18{,}980 \text{ days.}$$

This supracycle is the length of the Calendar Round. You can see that it is 73 Tzolkin cycles and 52 Vague Year cycles. Whenever the Calendar Round is to start over, there have been numerous special events in Mayan culture.

THE MAYAN CALENDAR CONTINUED: LONG COUNTS

There is another cycle that is an important part of the Mayan calendar. It is called the *Great Cycle* and is very long in years. Its definition is as follows: 20 *Kin* (days) = 1 *Uinal*, 18 *Uinals* = 1 *Tun*, 20 *Tun* = 1 *Katun*, 20 *Katuns* = 1 *Baktun*, 13 *Baktuns* = 1 Great Cycle. Notice how the values of 360 and 13 are incorporated into the structure. The Mayans did many calculations with values from the Great Cycle, denoting the dates obtained in what Mayan specialists call *Long Counts*. Many Long Counts appear on monuments built by the Mayans, called *stelae*. We will take a look at how Mayan Long Counts are described and some of the calculations with them. One thing we will want to determine is how long the Great Cycle is in the Western calendar, but that calculation will be easier after we look at some basic ideas of Long Counts.

First, we present some Mayan notation. The Mayans had several ways of expressing numbers. We will look at one form of notation of the Maya that is easy to use for working with Long Counts. In this notation, the Mayans used a dot, •, to denote values from one to four; a bar, | to denote groups of five; and a special symbol for zero: . Although the Mayans wrote Long Counts in various formats such as vertically or horizontally by rows (see Milbrath, 1999: 3–7), I will use a common Mayan format. In this format, the *Baktuns* appear at the top and the other units are read downward. The example below shows a Long Count from the archaeological site of Tikal in Guatemala. I have shown the bar and dot notation; a complete drawing of the stelae can be found in Milbrath (1999: plate 2).

Example 8.15 An example of a Mayan Long Count from Tikal.

Let us read the value. The values are written in descending order, but for calculations later, it is better for us to read it from bottom to top. There are two bars and two dots in the bottom symbol, so this is $5 + 5 + 2 = 12$, and the unit is *Kin* (days). The next level up has only one dot, so it is one *Uinal*. Next come three dots, which represent three *Tuns*. After that, there are fourteen *Katuns*. Finally, there are eight *Baktuns*.

For the rest of the examples of Long Counts, it may be helpful to write out the numerical values that occur in each of the levels. When combining

TABLE 8.4 Values of Various Levels of Long Counts

Level Name	Values
Pictun	$20^4 \times 18 \times 1$ to $20^4 \times 18 \times 19$
Baktun	$20^3 \times 18 \times 1$ to $20^3 \times 18 \times 19$
Katun	$20^2 \times 18 \times 1$ to $20^2 \times 18 \times 19$
Tun	20×18 to $20 \times 18 \times 19$
Uinal	20×1 to 20×17
Kin (day)	1 to 19

Long Counts, for example, when we accumulate a value of 20 in the bottom (*Kin*) level, that corresponds to one value (a dot) in the next level up, namely, the *Uinal* level. Symbolically, when we have four bars at the *Kin* level, it represents 20, which is one dot in the next level up. This rule applies to all the other levels *except* the third level (*Tun*). At this level, when we have a value of 18, we move up to the next level. This is because the *Tun* level represents $18 \times 20 = 360$, as opposed to 20×20, as would be the case for a purely base 20 number system, which the Mayans probably used for non-calendrical computations. Because combining Long Counts can lead to values beyond the *Baktun* level, another level called *Pictun* is added. Table 8.4 shows the details.

Before we get started on the next example, we can observe a few things about the Long Count. First, it is *positional.* That is, changing the order of the symbols changes the value of the Long Count. Second, the symbol of *zero* can, and is, used as a placeholder in the positional nature of the Long Count notation. Third, note the careful spacing of the symbols. If several dots are put close together, it could be confusing as to whether all dots should be at the same level or not. The Mayans were meticulous in writing this notation so as to avoid such confusion. Finally, you can see that the system can be extended beyond the *Pictun* level to arbitrarily large values.

As I alluded to earlier, we can determine the length of a Great Cycle. Recall that it is 13 *Baktun.* Table 8.2 tells us that this amounts to $20^3 \cdot 18 \cdot 13$, which is 1,872,000 days. If we divide this number by 365, we get approximately 5,129 years. The word "approximately" is interesting here for two reasons. One is that $20^3 \cdot 18 \cdot 13$ is not divisible by 365, so we do not obtain an integer value for that division. Another reason is that a Western calculation of Leap Year was not included in the calculation. We know the Maya calculated the lengths of Great Cycles without adjusting for what Western culture refers to as Leap Years, rather as a large count of days. At any rate, in the exercises you will have a chance to make such calculations and interpret them.

The Mayans calculated numerous calendar dates using Tzolkin, Vague Year, Long Count, and other cycles. We have already seen some such calculations using Tzolkin and Vague Year cycles. Now we would like to calculate a few combinations of Long Counts. You may wish to look back at Chapter 3 on number symbols to compare the Mayan symbols with, for example, the Egyptian symbols. Here is a reminder: Although I will describe the calculations using methods like carrying and borrowing values, there are other ways to view addition, subtraction, multiplication, and division. To subtract, for instance, one could arrange more convenient calculations based on the system of numeration. An example of this in the decimal system would be that to calculate 35 – 9, we could write it first as 35 – 5 = 30, then use our knowledge of base 10 to deduce that 4 less than 30 must be 26. That is, 35 – 9 = (35 – 5) – 4. So, keep in mind that *the Mayans did not necessarily use the methods presented here.*

Our first calculation will be to add. We add starting at the *Kin* (bottom) level, and work our way up. When necessary, we carry. I will write the bars horizontally with dots above them. This is another way in which the Mayans wrote the bar and dot notation (see Milbrath, 1999: 4).

Example 8.16 Add the following Long Counts.

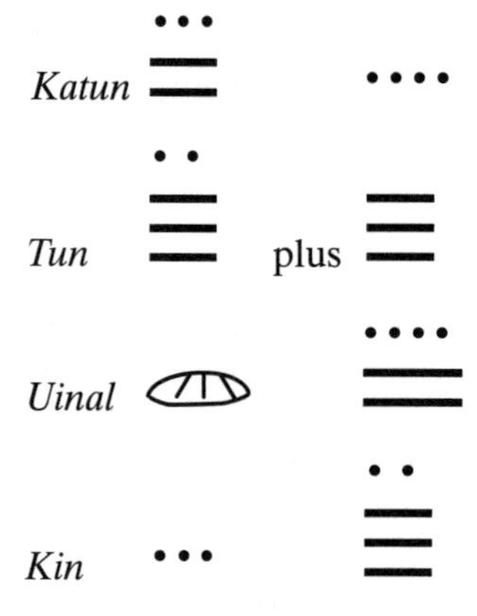

Below is a description of the steps involved, followed by diagrams of how the addition changes the bars and dots at each level. Note that you must read the diagram from bottom to top. At the *Kin* (bottom) level, we combine the three dots with the symbol to its right, of three bars with two dots, obtaining three bars and five dots. Five dots is another bar, giving us a total of four bars, which is 20. This means we carry one dot to the next level up and have zero left over. Hence, the bottom level of the sum will be the zero symbol. In the *Uinal* (second level from the bottom) level, we add the zero symbol to the symbol of two bars with four dots. Adding zero does not change the value, but we must remember that we carried a dot from the previous level, so we have five dots in all, making another bar. The value at this level will be three bars. Observe that if we had obtained a value of 18 or greater, we would have carried a dot to the next level. At the *Tun* level (third from the bottom), we combine the bars and dots to get a total of six bars and two dots. Separate four of the bars to make 20 and carry a dot to the next level, leaving two bars and two dots. At the *Katun* level (top), we have two bars, seven dots, plus one more dot that was carried from the level below, giving a total of eight dots. Write the eight dots as one bar and three dots. This gives us three bars and three dots at the top level (keep in mind the dot that was carried up from the *Tun* level). Now look at the diagrams.

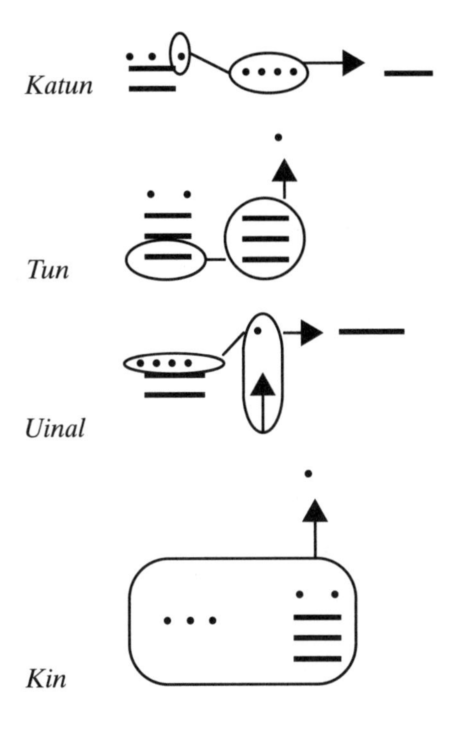

Below is the final Long Count.

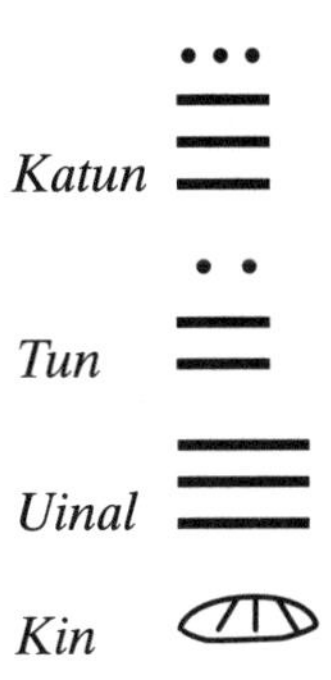

Here is another example.

Example 8.17 Add the following Long Counts.

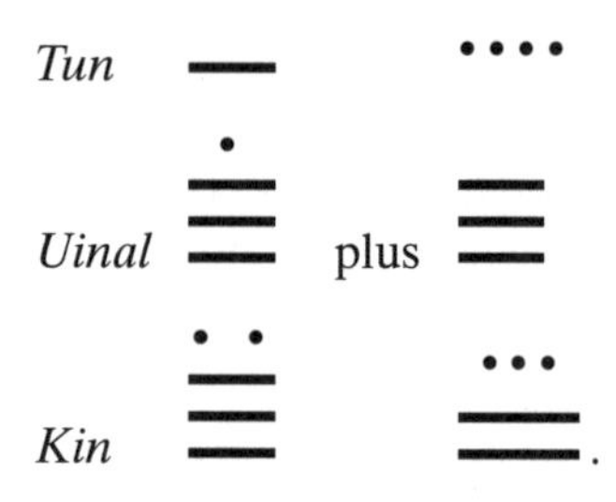

Starting again at the *Kin* level, there are five bars and five dots for a total of six bars. Separate four and carry a dot to the *Uinal* level, leaving two bars at the *Kin* level. Next, there are six bars, one dot, and the carried dot at the *Uinal* level. When we have 18 at the *Uinal* level we carry, so separate three bars and three dots to carry to the next level, as shown below.

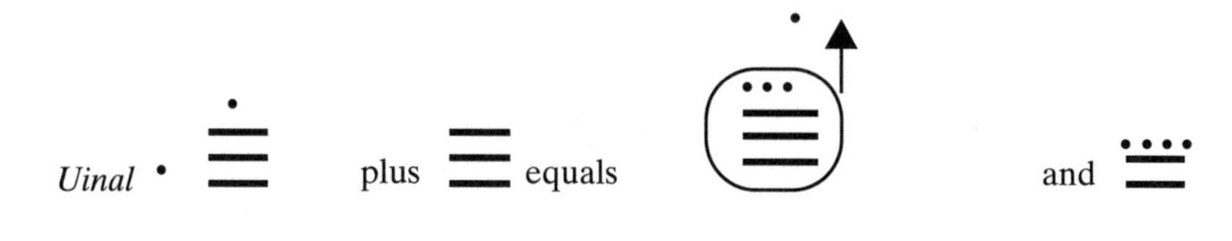

Thus, we carry the part to the next level as a dot, leaving the two bars and four dots. For the *Tun* level, we have one bar, four dots, and the carried dot, giving us five dots in all, and we write the five dots as a bar. The final sum is shown below.

Tun

Uinal

Kin

Next is subtraction.

Example 8.18 Subtract the following Long Counts.

Tun

Uinal minus

Kin

Starting at the *Kin* (bottom) level, we do not have enough value to subtract, so we *borrow* from the *Uinal* (second from the bottom) level. This will leave two dots in the *Uinal* level, and we add four bars (20) to the *Kin* level, all in the left-hand number. A diagram of the borrowing is shown below.

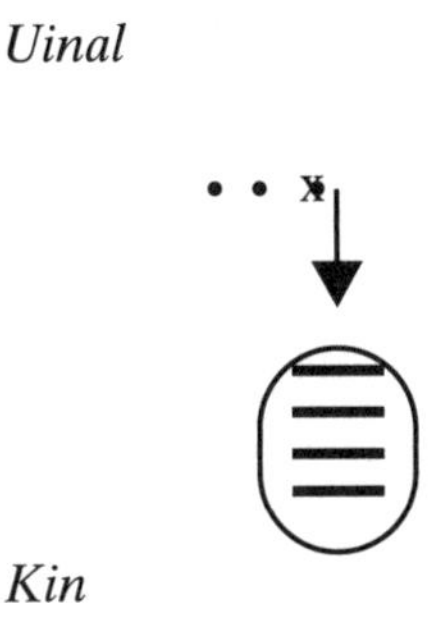

Subtracting in the *Kin* level now looks like this:

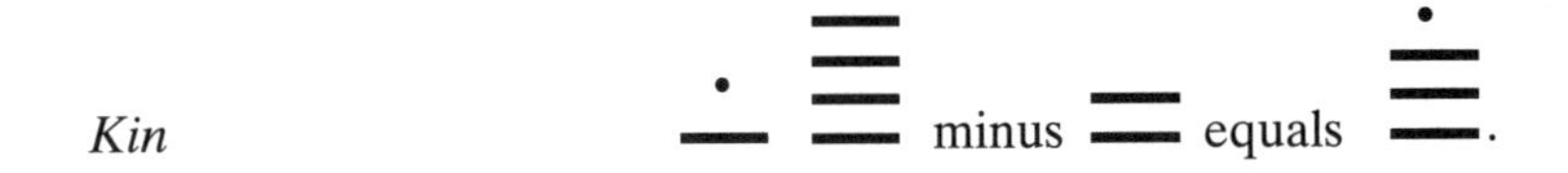

Notice that we can just cross out the two bars from the second number, corresponding to two bars from the first number.

In the *Uinal* level, we have two dots left, and that is not enough to subtract the three bars to the right, so we borrow from the *Tun* level, keeping in mind that it is a value of 18 that is borrowed, not 20. The *Uinal* level operation looks like what is shown below.

Uinal minus equals.

For the *Tun* level, we have just one dot left and we subtract zero from it, leaving just the one dot. The final value of the subtraction is shown below.

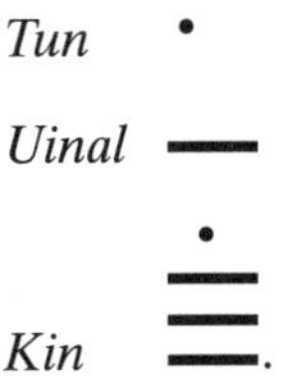

Here is another example of subtraction.

Example 8.19 Subtract the following Long Counts.

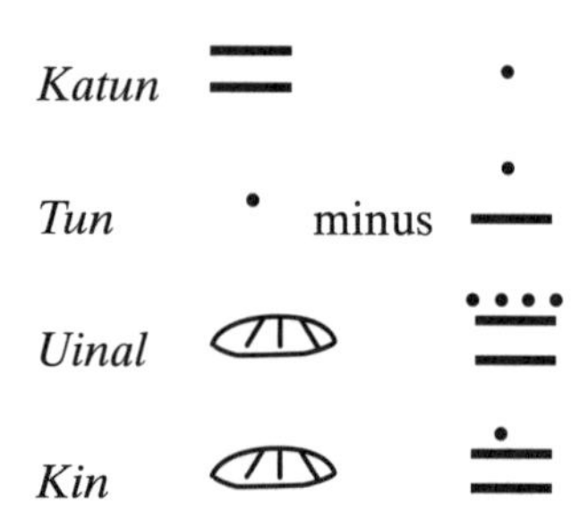

As usual, we start at the *Kin* level. Because there is a zero in the left value, we must borrow, and all the way to the *Tun* level because of the zero in the *Uinal* level. Compare this to when we subtract something like 3001 – 25 in the decimal system, in which we must borrow from several levels because of the zeros. Thus, we borrow from the *Tun* level. However, recall that when we carry from *Uinal* to *Tun* it is a value of 18 and not 20, so to borrow from *Tun* to *Uinal* it is a value of 18 that is "sent down." The diagram for the borrowing from *Tun* to *Uinal* is shown below.

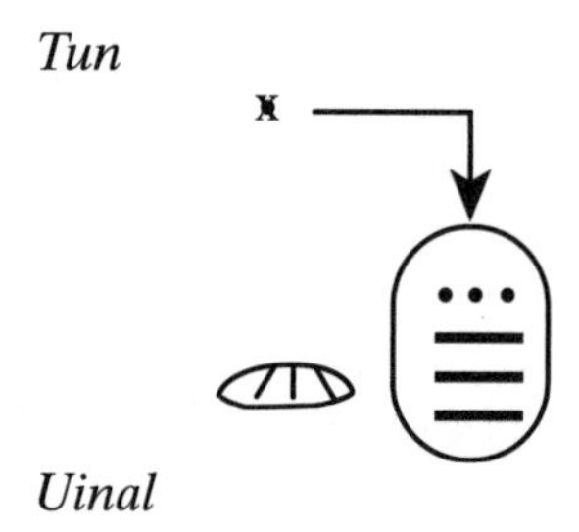

We now have a value of 18 in the *Uinal* level and a zero in the *Tun* level of the first number. We must borrow from the *Uinal* level to the *Kin* level, so from those 18 we take out one dot, remembering that this time, what gets borrowed is a value of 20 instead of 18. See below.

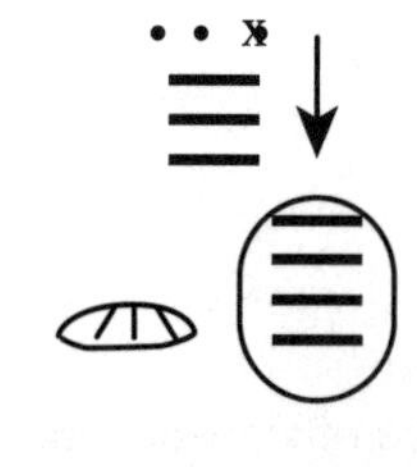

Kin

We are almost ready to finish the subtraction, but thinking back to the *Tun* level, we are left with zero at that level, and will have to subtract one bar and one dot (value of 6) that appears in the second number. Because we will have to borrow at that level, let us do that now. We borrow from the *Katun* level. Recall that we have a value of only 18 between the *Uinal* and *Tun* levels, and 20 between all other levels, so the value we borrow here is 20. We split one of the two bars from the *Katun* level into five dots and then take out one to put in the *Tun* level as 20. Below is the diagram.

Katun

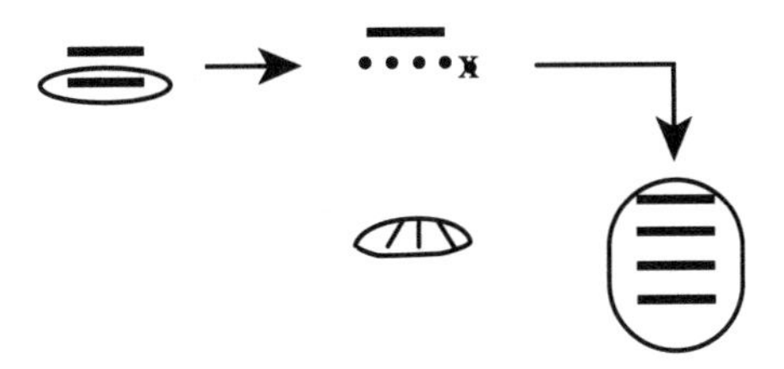

Tun

With all the borrowing finished, let us rewrite the subtraction expression again, this time with the adjusted symbols in the first number from that borrowing.

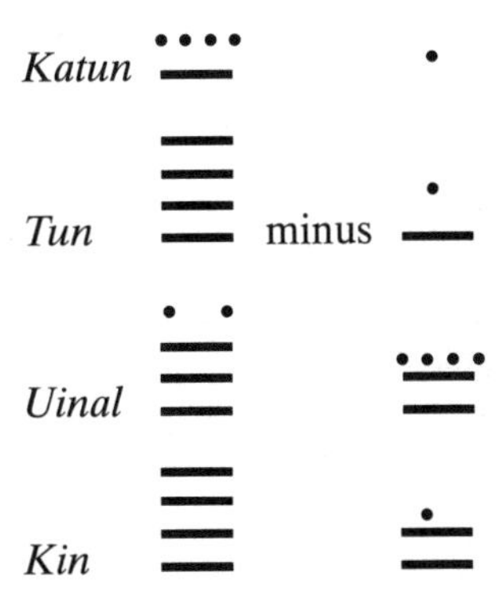

To perform the subtraction is now relatively easy. At the *Kin* level there are four bars in the first number and two in the second, so delete two bars from the first number. Then, there are two bars left and we must subtract one dot from those two bars. So, write one of the bars as five dots, and delete one that corresponds to the one dot from the second number. What remains (which is what subtraction is about, right?) at the *Kin* level is one bar and four dots. The subtraction process looks like this:

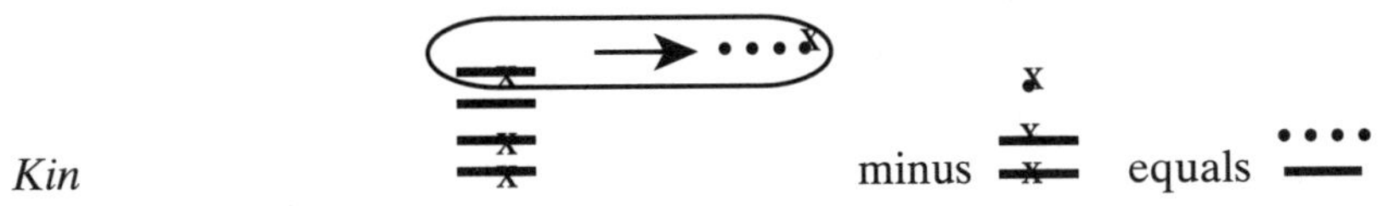

At the *Uinal* level we delete two bars from the first number because there are two bars in the second number. Then we write the remaining bar as five dots. From that point, we can delete corresponding values as we did for the *Kin* level. Continuing the process of crossing out corresponding symbols at each level, and replacing five dots with a bar (as occurs in the *Uinal* level), we obtain the final value of the subtraction, shown below.

Katun •••
—

Tun ••••
—
—

Uinal •••

Kin ••••
—

This completes our short look at Long Counts. We could continue with more operations such as multiplication, division, and so on; however, such an excursion would make this chapter too long.

A few final comments are in order as we finish this chapter. The above description of the Mayan calendar gives you an idea of how their calendar worked. There are certainly other calculations that we could try, such as combining Long Counts with other cycles like the Tzolkin or Vague Year. In fact, the Mayans did these and many more calculations. Moreover, they solved sophisticated problems related to their calendar that are beyond the level of discussion here. Good descriptions of some of the calculations can be found in Closs (1986: chapter 11), Milbrath (1999: introduction and appendix 3), and Ascher (2002: chapter 3).

We have seen that calendars can be based on natural or physical cycles or on culturally determined cycles, or have combinations of these types of cycles. How the calendar functions in a culture depends on how that particular culture uses the calendar in its activities. The Nuu-chah-nulth calendar represents an example of a tool for keeping track of cyclical changes in the environment, and the Mayan calendar represents a tool for keeping track of cultural activities, astronomical cycles, and agricultural cycles. While very different in styles and structures, each type of calendar was useful to the corresponding culture.

EXERCISES

Short Answer

8.1 **(a)** What is the difference between a lunar cycle and a month? Which, if any, of these is culturally determined and which, if any, is physically determined? Explain.

(b) Is a week culturally determined or physically determined?

8.2 Does a calendar have to start on the same day every year? If not, can you describe some examples of calendars that do not start on the same day each year?

8.3 **(a)** What is a solstice? How is it observed (without special astronomical equipment)?

(b) Is a solstice physically determined or culturally determined? Explain.

8.4 **(a)** What is a leap year? How do we know when a specific year will be a leap year? Is this the only way to make the adjustment? Explain.

(b) Is a leap year physically determined or culturally determined? Explain.

8.5 **(a)** What is an equinox? How is it observed (without special astronomical equipment)?

(b) Is an equinox physically determined or culturally determined? Explain.

(c) Does an equinox occur on the same day for all parts of the earth?

8.6 Is the degree measurement on a circle physically determined or culturally determined? Explain.

8.7 What happens in the Nuu-chah-nulth calendar when it is a leap year in the Western calendar? How would the Nuu-chah-nulth include a leap year into their calendar? Explain your responses.

8.8 **(a)** Are the Mayan cycles physically determined or culturally determined? Explain.

(b) Which days, if any, have special ritual significance in the Mayan calendar system? Explain.

(c) Which days, if any, have special ritual significance in the Western calendar system (not including holidays, vacation, weekends)? Explain.

8.9 Consider the following excerpt from Moziño's descriptions of the Nuu-chah-nulth calendar (as described in Folan, 1986):

> [O]utsiders considered Yuquot Nootkan chronology to be obscure, either because the Europeans had difficulty understanding it or because the Yuquot Nootkans were careless about arranging their "calendars.". . . The chiefs decided when days should be added to any particular period and when the next period should begin, and because these decisions were based on variable features, uncertainty regarding Yuquot Nootkan calculations of time always existed. (pp. 94–96)

Explain how these statements indicate a Western view of mathematics, indicating which Western views are being assumed in the statement.

Refer to Chapter 1 for descriptions of the different Western views of mathematics.

Calculations

8.10 Calculate the least common multiples and supracycles.

(a) $lcm(54,114)$.

(b) $lcm(25,64,17)$.

(c) Suppose one cycle of a calendar has 36 days and another cycle has 60 days. If the two cycles coincide today and run independently of each other, determine how many days it will be until the two cycles coincide.

(d) Determine when the two cycles in part (c) will coincide if the cycle of 60 days must finish before the 36-cycle advances.

8.11 Suppose we were to change our calendar so that it only has weeks and no months. Determine and explain the following:

(a) How many weeks would there be in each year? Explain your calculation.

(b) What adjustments, if any, would have to be made during the course of many years?

(c) If the first day of the year is on a Tuesday, then on which day of the week is the 255th day? What about the 171st day? Explain your calculation in terms of modulo 7.

8.12 Using the descriptions of the Nuu-chah-nulth cycles from Table 8.1, determine and explain the following:

(a) Which, if any, of the cycles are physically determined?

(b) Which, if any, of the cycles are culturally determined?

8.13 For the Mayan calendar consider the cycle of 360 days as a separate cycle. Determine when all three of the 365-, 360-, and 260-day cycles coincide. Write your solution in units of days, Tzolkins, Vague Years, and solar years.

8.14 The next Great Cycle is scheduled to end on December 21, 2012. Using the ideas from calculating the number of days in a Great Cycle (see shortly after Table 8.2), calculate:

(a) In what year (Western calendar) did the current Great Cycle begin?

(b) What year do you get if you include a simple (i.e., before the Gregorian Western calendar) calculation of leap year, by adding add one day every four years?

(c) What year do you get if you use a value of 365.2422 for the number of days in a tropical (solar) year?

(d) Write about or discuss your interpretations of the calculations from parts (a)–(c).

(e) Assuming that on December 21, 2012, the current Great Cycle will end, determine the month, day, and year (Western calendar) when the next Great Cycle will end. Note that you will need to find a calendar calculator (check the Internet) for determining days, months, and years well into the future.

(f) Based on your calculation from part (e), and keeping in mind that the current Great Cycle is scheduled to end on an astronomically significant day (it's a fall solstice), would you need to adjust the date you determined? If so, how? Explain your interpretations of this calculation of the next Great Cycle within the cultural context of how the Mayans would view this aspect of their calendar.

8.15 For each of the dates below, determine the new Tzolkin date:

(a) 631 days after 12 *Caban*. **(b)** 444 days before 9 *Chuen*.

(c) Two Tzolkin cycles before 5 *Eb*.

8.16 Suppose a Tzolkin date has equal indices; that is, it is of the form ${}_jT_j$.

(a) Explain how to determine how many days it will be until both indices coincide again.

(b) Verify your solution from (a) for the Tzolkin date ${}_9T_9$.

8.17 Calculate the Calendar Round dates (i.e., $({}_iT_j, {}_mV_l)$) for the following. Use Tables 8.2 and 8.3 as necessary.

(a) 873 days past the date (13 *Manik*, 19 *Zac*).

(b) 740 days before the date (13 *Manik*, 19 *Zac*).

(c) 1000 days before the date (9 *Caban*, 16 *Kayab*).

8.18 Suppose today has a Calendar Round date of (13 *Imix*, 5 *Zip*), as $({}_{13}T_1, {}_5V_3)$. Determine how many days it will be until the first of the extra five days.

8.19 The Mayans knew that Venus, the sun, and Earth were in alignment approximately every 584 days. If they are aligned on the Vague Year date of 17 *Mac*, determine the Vague Year date that represents the *last time* they were aligned.

8.20 Suppose there is a day for which *all four* of the indices in the Calendar Round are the same. When, if ever, will a Calendar Round date have all four indices the same again? If it will not occur again, explain why. If it will, explain how one would determine how many days it would be until all four are the same. *Note*: For simplicity, start with $i = j = m = l = 1$.

For Exercises 8.21–8.24, perform the indicated operations on the Long Counts. Explain your intermediate steps! Be careful with spacing. The Long Counts are from actual Mayan locations; however, keep in mind that we do not know if the Mayans calculated particular operations in the same way as presented here.

8.21 The first value is from Copán, Honduras, the second from Yaxchilán, Mexico.

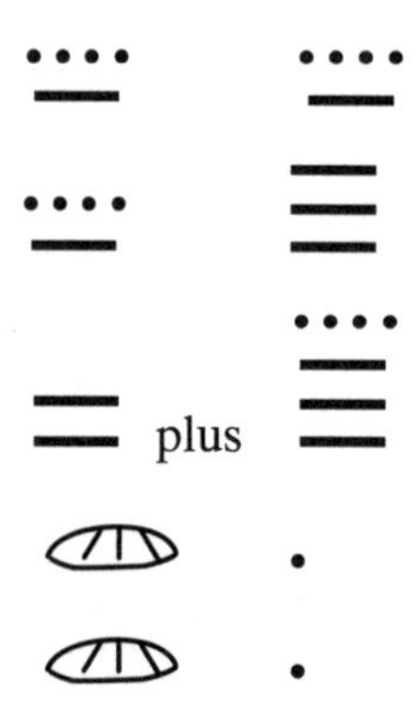

8.22 The first value is from Palenque, Mexico, the second from Tikal, Guatemala.

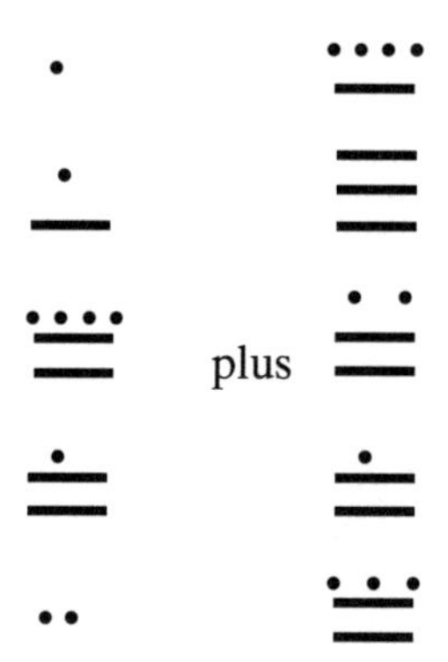

8.23 Both values are from Tikal, Guatemala.

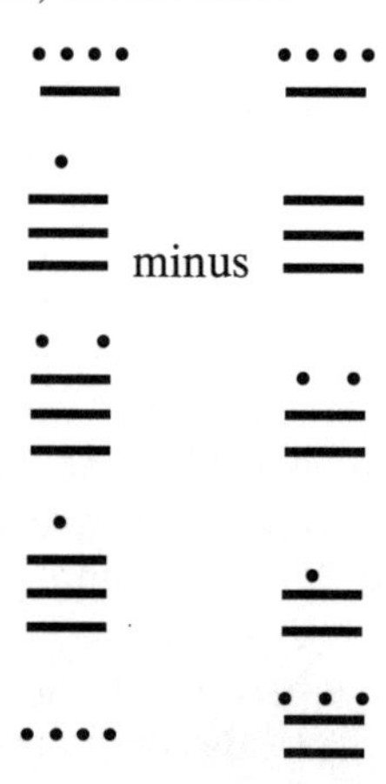

8.24 The first value is from Ixlu, Guatemala, the second from Palenque, Mexico.

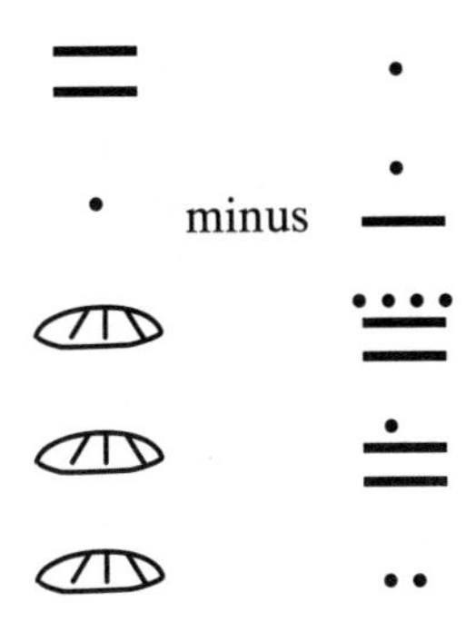

Exercises 8.25–8.27 below assume some familiarity with doing mathematical proofs. These properties may have been known by the Mayans, certainly those of Exercise 8.26.

8.25* Prove that if b is any *odd* integer, then $b^2 \equiv 1$ mod(8). *Hint*: $b = 2m + 1$ (m is an integer). Calculate b^2 and consider both of the cases: When m is even, and when m is odd.

8.26* Prove the following properties of modular arithmetic (all variables are integers):

(a) $a \equiv a \pmod{n}$ is always true.

(b) If $a \equiv b \pmod{n}$, then $b \equiv a \pmod{n}$.

(c) If $a \equiv b \pmod{n}$, and $b \equiv d \pmod{n}$, then $a \equiv d \pmod{n}$.

(d) If $a \equiv b \pmod{n}$, then for any $c, a + c \equiv (b + c) \pmod{n}$ and $ac \equiv (bc) \pmod{n}$.

8.27* **(a)** Prove that for any integer a, $a^3 \equiv 1 \pmod{n}$, where n is divisible by $(a - 1)$.

(b) Verify the result of (a) for the value of $a = 177$, and also for the value of $a = -91$.

Essay and Discussion

8.28 Construct a calendar like that of the Nuu-chah-nulth, based on your activities throughout the year. For example, there could be a "final exam" cycle, a "summer vacation" cycle, and so on. Your calendar should have at least eight cycles, and must cover an entire year. Be sure to explain how cycles would be adjusted, and describe an example of how your adjustment scheme is applied to one or more of your cycles. You may start your calendar cycles near January 1st, but you are not obliged to start your year then. Write about or discuss the details of your calendar.

8.29 Write about or discuss your explanations of why the Mayans created a calendar system that has many cycles in it. Include descriptions of which mathematical concepts the Mayans probably knew, such as congruences and least common multiples.

8.30 Compare the Nuu-chah-nulth calendar with the Mayan calendar. Include comparisons between the types of cycles used in each calendar, and what each calendar was used for. Include brief descriptions of the mathematics used in each calendar.

8.31 Write about or discuss about interpretations of the Nuu-chah-nulth calendar. Describe what the calendar indicates about how the Nuu-chah-nulth understand time and number. Indicate how the Nuu-chah-nulth interpretations of time and number differ from how time and number are viewed in Western culture, and explain what those differences imply regarding understanding Nuu-chah-nulth mathematics. You may wish to page through Chapter 1.

8.32 Suppose there is a plan for some people to travel for an extended time to Mars. Write about or discuss your investigation in which you describe how to design a calendar based on the Martian cycles. Here is some information:

- One Martian day is: 24.6 Earth hours (1 Mars day is sometimes called 1 sol).
- One Martian year: 1.88 Earth years = 669.908 Martian days.
- Mars has two small moons that rotate around it as follows: Phobos: 11.12 Earth hours, Deimos: 131 Earth hours. These are average moonrise times (the time it takes to rise in the east and set in the west).

You will have to specify certain conditions according to which you design your calendar, such as whether you use Earth hours and days or Martian hours and days. For Martian hours, you will have to define them in terms of units in which you divide a day (it does not have to be 24). Also, you will have to explain whether or not you will adjust for long-term accumulations in the Martian year. It could be like a leap year calculation, but need not be. For this to be interesting, it would be best to try to design the calendar based as much as possible on Martian cycles and not Earth cycles.

8.33 Design a calendar as in Exercise 8.32 above for another planet other than Earth or Mars. You will have to look up the relevant information about the planet such as how long it takes to rotate about the sun and details of its moons (if any). If the planet has several moons (e.g.,

Jupiter), you will have to figure out which ones to include in your cycles and how the cycles will be measured. If the planet has no moons (e.g., Venus) you will have to determine some way to measure cycles that does not include lunar cycles.

8.34 Some people have proposed that long ago that the Egyptians crossed the Atlantic Ocean and landed in an area of the Western Hemisphere that was Mayan territory at that time. Presumably, the Egyptians taught the Mayans how to construct their pyramidal structures. Write about or discuss an analysis of this possibility from the point of view of cultural mathematics. You may wish to look back at Chapter 3 where Egyptian symbols were discussed. Specifically, assuming that years of interaction between the Mayans and the Egyptians would manifest itself in the mathematics of both cultures, address the following in your considerations.

(a) Is there any evidence of a connection between the counting base of the Egyptians and the counting base of the Mayans? Is there any evidence of a connection between mathematical symbols of the Mayans and Egyptian mathematical symbols?

(b) Is there any connection between the positional and nonpositional natures of the two mathematical systems?

(c) How would your responses to parts (a) and (b) explain the daily process of the Egyptians teaching the Mayans how to construct pyramids, buildings, and other structures?

(d) Describe your conclusions regarding how likely it is that the Egyptians really did teach the Mayans how to build their structures. Remember, your conclusions should be based on the mathematical aspects described in the previous parts of this question.

(e) Describe the *cultural* aspects of the possibility of the Egyptians coming to Mayan territory to teach them about construction. Include considerations such as how the Mayans would view this kind of conclusion, and how the Egyptians would view it as well.

Additional Comments

- Although I have highlighted some parts of Folan (1986) as examples of Western views of mathematics, I should mention that in general his article does not represent a negative view of Nuu-chah-nulth (Nootkan) calendars and mathematics. In his descriptions of Nootkan number expression, for example, Folan describes how the Nuu-chah-nulth perceive numbers and quantities in ways that are different from how such things are interpreted in Western culture. Moreover, he mentions several times that early Europeans who were in contact with the Nuu-chah-nulth erroneously assumed decimal counting (the Nuu-chah-nulth count in base 20).

- Mayan specialists have a notation for Long Counts in which each unit is separated by a period, and the units are listed in order from right to left. For example, a Long Count of 19 *Kin*, 16 *Uinal*, 2 *Tun*, 7 *Katun*, 14 *Pictun* would be written as 14.0.7.2.16.19. Notice that the *Baktun* value is zero, and the values for *Uinal* are from 0 to 17 (because of moving to the next level at 18 *Uinal*).

PART II

CASE STUDIES

Motivational Questions: Before you begin reading Part II of this book, take some time to think about your responses to the following questions.

- *Can you summarize the main points of cultural mathematics that we have seen so far?*
- *What would be some obstacles to understanding the mathematics of another culture?*
- *How could we overcome the obstacles from the previous question?*

The first goal is to summarize the main points that we have seen so far in Part I. After that, we want to get ready for the two chapters that follow on the Otomies and the Incas. However, in order to make a complete study of the mathematics of another culture, we must determine where to look for information on mathematics of another culture, including advantages and disadvantages of the resources. We start with a summary of what we have seen so far.

Up to this point, we have discussed many examples of mathematics from many distinct cultures. Now is a good time to list the main points that bring together what we have looked at.

- Mathematics is much more than formulas, equations, and theorems. It is a way of thinking that can be described in general terms of categories, such as with Bishop's six.

Introduction to Cultural Mathematics: With Case Studies in the Otomies and Incas, First Edition.
Thomas E. Gilsdorf.

- People from distinct cultures understand, interpret, and express mathematics in different ways.
- Activities that at first do not appear to involve mathematics often do involve mathematics, sometimes mathematics that is quite sophisticated by Western standards.
- Cultural tendencies can be both subtle and strong. We must work hard at recognizing our own cultural tendencies and distinguishing between how we perceive mathematics and how mathematics is perceived by other cultures.
- Because the context of our work is to observe other cultures from the point of view of Western culture, it behooves us to list again the Western views of mathematics that we discussed in Chapter 1:
  - An emphasis on number and precision.
  - Assumptions that other cultures follow a path of development similar to how Western culture has developed.
  - A preference for cultures that have developed a system of writing.
  - Assumptions about gender roles and mathematics.

WHAT INFORMATION IS KNOWN ABOUT CULTURAL MATHEMATICS?

One thing you will find, and that will come up in Part II, is that for many cultural groups we often do not have a complete picture of that culture's mathematics. It might be, for example, that only some number expressions and some basic information about calendars is known about the mathematics of a particular culture. For the case studies we will look at, we will not be able to describe some aspects for which information is not yet known. If you decide to learn about cultural mathematics (or ethnomathematics, as it is also called), there are plenty of cultures that you could investigate with respect to mathematics!

RESOURCES FOR LEARNING ABOUT CULTURAL MATHEMATICS

This section has a general description of where to look for information about cultural mathematics. I have tried to describe each resource in terms of both advantages and disadvantages.

Resource: *Descriptions of one culture by another culture, often from a long time ago.* Many such descriptions are part of an explanation of one culture by people of another culture, which was later transcribed into writing. The description could be by early explorers or settlers who came upon an area inhabited by a culture with which they were not familiar. Most such descriptions that you are likely to come across are descriptions by people of Western culture. There are both advantages and disadvantages to this

Advantages: One advantage can be that the information is described during an early contact with the culture in question, before outside influences have taken place. Such descriptions often represent the earliest known contact with other cultures. They may contain accurate descriptions of the culture.

Another advantage can be that the writer of the information may have spent a lot of time with the people of that culture, and hence has accurate and less culturally biased views.

Disadvantages: Until fairly recently, Western researchers did not take into account the cultural tendencies of the person doing the research. Thus, many descriptions from long ago are culturally biased. Often there are assumptions that people of the other culture did not develop any mathematics. Hence, important mathematical information can be ignored in the research.

Another disadvantage is that frequently the person doing the research had ulterior motives for the research. For example, church workers in the Americas described many cultures during the 1500s and 1600s, but their descriptions were usually in the context of wanting to convert the people of the culture they studied.

A third disadvantage of old descriptions of other cultures is a lack of mathematical expertise of the researchers. There are artifacts like the Inca *quipu* (that we will see in Chapter 10) that require an understanding of rather sophisticated mathematical concepts in order to describe them. Because many of the people who studied other cultures did not have a deep mathematical background or were not intentionally looking for mathematical concepts, important mathematical information can be overlooked. Another consideration is that, for example, in the 1500s when many cultural descriptions were written, good understandings of mathematical concepts like modular arithmetic (which we saw in Chapter 8 is important for understanding calendars) was not well understood by Western explorers who traveled to the Americas, parts of Africa, or other places.

Yet another disadvantage of descriptions by one culture of another culture is problems resulting from language translation. Languages that are completely new to a researcher represent significant obstacles to understanding information. Sometimes information gets mistranslated, causing confusion.

Perhaps at this moment you are thinking that information from older resources is never to be trusted, but in my experience much of such information is quite good. The point is rather that you need to be aware of the less obvious considerations discussed here as disadvantages, so as to be able to grasp the information for which you are searching. Now, let us look at another resource.

Resource: *Recent descriptions by anthropologists, archaeologists, linguists, and historians, or interviews with such persons.*

Advantages: These disciplines have changed much over time, and recent work by anthropologists, archaeologists, linguists, and historians shows an approach in which one is quite careful to distinguish between cultures and cultural tendencies. Moreover, there are rather sophisticated techniques for obtaining and analyzing information that could be very useful. In general this is an excellent resource. Nevertheless, there is one small disadvantage.

Disadvantage: The main consideration here is that for most anthropologists, archaeologists, linguists, and historians, a careful study of the mathematics of a culture is usually not a high priority of their research. Overall, keep in mind that part of a study of the mathematics of a cultural group must include significant information about the culture in general, and people with research skills in areas such as anthropology can be of great help, even if the mathematics of the culture has not been their main focus.

Resource: *Interviews with people of the culture you want to study.* Firsthand information about a culture can be very valuable, especially because it comes directly from the culture.

Advantages: The most important advantage is that you could obtain good information that might not be available in any other form. When it comes to mathematics, you will find that there is surprisingly little information in the form of books and journal articles about mathematics for many cultures around the world. So, a direct interview might be a very important resource. On the other hand, there are disadvantages.

Disadvantages: The term "disadvantage" might not be the best choice here, because the main point about interviews is that they are more complicated than you might first expect. For one thing, you have to make sure the questions you ask do not inadvertently steer responders toward a particular conclusion. The wording of the questions can be important! Another consideration is that the context of your interviews should be that the people you talk to feel comfortable with the process and are open to responding to the kinds of questions you want to ask. Indeed, some people may feel uncomfortable answering questions that you might think are harmless. Also, in some cultures certain kinds of information that is considered sacred or otherwise important is not to be discussed with people from outside the culture. It is up to you to respect such restrictions. Third, because you will be working with humans directly, there are numerous considerations regarding human subjects in research. For one thing, many countries have specific laws and/or guidelines as to what is considered appropriate when it comes to activities such as interviews. In the United States, the Institutional Review Board (IRB) carefully reviews research that involves human subjects. My general advice is that you should feel motivated to conduct the interviews, but be prepared to go through a very careful process to get it set up. If you are considering interviewing some people for the first time, my suggestion would be that you first contact someone who has conducted research using interviews to find out the basic procedure.

We are ready to start our first case study, the Otomies.

9

HÑÄHÑU MATH: THE OTOMIES

Motivational Questions: Before you begin reading this chapter, take some time to think about your responses to the following questions.

- *The Otomies are a major cultural group of present-day Mexico, but the Otomi language is distinct from the languages of the Aztecs and Maya. What does this language difference say about the Otomi culture compared with the Aztec and Maya cultures?*
- *The Otomies were already established as a culture in central Mexico when the Aztecs migrated there. What does this imply about the possible cultural and mathematical interactions that have taken place between the Otomies and Aztecs?*
- *Several cultural groups that have lived in central Mexico, including the Otomies, have similar counting systems. Why do you think the counting systems are similar?*
- *When we study the Otomi artwork in this chapter, we will look for symmetry properties. What kinds of conclusions could we draw about Otomi culture that might be revealed by such symmetry properties?*
- *There is a description of the Otomi calendar that includes a statement to the effect that the Otomies added six hours to their calendar each year in order to account for a leap year. What do you think of that statement? Are there other ways the Otomies could have accounted for a leap year?*

Introduction to Cultural Mathematics: With Case Studies in the Otomies and Incas, First Edition.
Thomas E. Gilsdorf.

INTRODUCTION

Before starting this chapter, it is important to make some general observations. First, you will find that we will not describe all the aspects of cultural mathematics that were discussed in Chapters 1–8. This is because, in this case, we do not have the information on every aspect of cultural mathematics of the Otomi culture. Another comment to be made is that we will see that for some aspects of the cultural mathematics of the Otomies, we must take into consideration the influence and interaction on Otomi culture that has occurred from the Aztec and Spanish cultures. These interactions blur the lines between what is Otomi culture and what is from another culture. It is worth recalling the general term of *Mesoamerica*, which refers to the geographic area ranging approximately from present-day central Mexico to the southern border of Honduras. The Otomies are a Mesoamerican cultural group.

This chapter is about the mathematics of the cultural group known as the Otomies, of present-day central Mexico. In particular, I discuss the Otomi number system and a comparison of that system with Aztec counting, Otomi art and decoration, mathematical symbols that appear in some Mesoamerican codices, and the Otomi calendar.

WHO ARE THE HÑÄHÑU?

The cultural group we are going to consider is generally referred to as the Otomies (pronounced "oh-toh-MEES"). The name Hñähñu ("hny-EH-hnyu," with a nasalized "h") is one of the names by which people of this culture refer to themselves. In general, people of Hñähñu culture use a variant of this word when referring to themselves with other members of the same culture, but use the word *Otomi* in communications in Spanish or with people from outside the culture. The term *Otomi* has been used in a derogatory way since before the arrival of the Spanish. Nevertheless, there is some consensus among anthropologists who study Hñähñu culture that it is better to use the word *Otomi*, presumably in a positive way, so as to work toward creating a more positive image of the people who are often referred to with this word. A discussion of the meanings of the terms *Hñähñu* and *Otomi*, and the ways in which the Otomies have been portrayed over the years, can be found in Wright (1997 and 2005: 19). I will use the term *Otomi* and hope that this chapter will serve as a positive description of the people, as viewed through a study of their mathematics. Our first goal will be to get to know who the Otomi are.

The Otomies are an indigenous group of what is now central Mexico. The modern-day territory in which significant numbers of Otomies live and practice some form of their traditional culture includes large parts of the Mexican states of Hidalgo and Mexico, with smaller areas in the states of Puebla, Guanajuato, Michoacán, Querétero, and Veracruz. There is also a significant number of Otomies in Mexico City. Over the last several centuries the Otomies have migrated because of issues such as conflicts with the Aztecs and the arrival of the Spanish in 1519.

It turns out that the Otomies represent one of the largest and oldest indigenous populations of what is now Mexico. According to Yolanda Lastra (2006: 25), the number of speakers of Otomi (including those who also speak Spanish) as of 2000 was about 292,000. This gives us a rough idea of the number of people who practice Otomi culture. David Wright (1997: 439–441) explains that there is significant evidence to conclude that the Otomies represent one of the oldest cultural groups of what is now central Mexico. Moreover, the Otomi language is part of the Otomangue family (see Soustelle, 1937, or Lastra, 2006). The Otomangue language family is neither Yuto-Aztecan nor Mayan. Think about the motivational question about language. Our general statement is: The Otomies represent a very old and very large cultural group, and one that is distinct from the more well-known Aztec and Mayan cultures. Another point to make is that when the Aztecs migrated to what is now Mexico City, the Otomies were already an established culture in the area. Let us take these facts into consideration and look more closely at the Otomi-Aztec history.

The Otomies and Aztecs have lived in the same general area since the Aztecs migrated to what is now central Mexico in the 1100s. The interaction between these two groups has not always been positive. At first, the Aztecs were a small group that probably had little influence over pre-established groups such as the Otomies. Later on, by the 1400s, the Aztecs had created a large empire that controlled many cultural groups and much territory in the area, including most of the Otomi territory. Many of the groups controlled by the Aztecs were resentful of being controlled, leading to conflicts with the Aztecs. The Otomies were one such group. In 1519, Hernán Cortés arrived in what is now Mexico City and eventually took control of the Aztecs. Some of the groups that had resented the Aztecs supported Cortés in taking over the Aztec capital (where Mexico City is now). Many Otomies took part in this conflict as well. Thus, for over 400 years, the Otomies and Aztecs shared the same geographical area and were in frequent contact, at times in conflict, with each other. Because of this long-term interaction, the mathematics of the Otomies and Aztecs have blurred together such that we are not always able to determine which of these cultures (or another culture of the area) was the first to use a particular mathematical idea. Understanding the mathematics of the Otomies will require us to understand some basic facts about the mathematics of the Aztecs. Our first topic is Otomi numeration.

OTOMI NUMERATION

We will look at a list of Otomi number expressions and determine the mathematical constructions that go with it, as we did in Chapter 2 with other cultures. After that, we can make a comparison between Otomi and Aztec number

TABLE 9.1 Otomi Number Words

Number	Otomi Expression	Number	Otomi Expression
1	*ˀna*	35	*ˀnate ma- ˀrætˀa ma- kïtˀa*
2	*yoho*	39	*ˀnate ma- ˀrætˀa ma- gïto*
3	*hñu*	40	*yote*
4	*goho*	41	*yote ma- ˀra*
5	*kïtˀa*	50	*yote ma- ˀrætˀa*
6	*ˀrato*	60	*hñu ˀrate*
7	*yoto*	63	*hñu ˀrate ma- hñu*
8	*hñato*	85	*goho ˀrate ma- kïtˀa*
9	*gïto*	100	*ˀna nthebe*
10	*ˀrætˀa*	200	*yo nthebe*
11	*ˀrætˀa ma- ˀra*	500	*kïtˀa nthebe*
12	*ˀrætˀa ma- yoho*	1,000	*ˀna mahua¸hi*
13	*ˀrætˀa ma- hñu*	20,000	*ˀnate mahua¸hi*
16	*ˀrætˀa ma- ˀrato*	50,000	*yo ˀnate ma- ˀrætˀa mahua¸hi*
17	*ˀrætˀa ma- yoto*	65,789	*hñu ˀrate ma- kïtˀa mahua¸hi ne yoto nthebe ne goho ˀrate ma- gïto*
20	*ˀnate*		
21	*ˀnate ma- ˀra*		
22	*ˀnate ma- yoho*	100,000	*ˀna nthebe nehua¸hi*
23	*ˀnate ma- hñu*	800,000	*hñato nthebe nehua¸hi*
30	*ˀnate ma- ˀrætˀa*	1,000,000	*ˀna mahua¸hi de ga mahua¸hi*

expressions and see if there are any conclusions that come from that comparison. The Otomi number words in Table 9.1 come mainly from the dialect of Otomi spoken in the Mezquital Valley of the state of Hidalgo. I have taken the information for these number words, including the linguistic notation that appears, from Barriga Puente (1998: 229) and *Luces Contemporaneas* (1979) (see also Thomas, 1897; Merrifield, 1968; Acuña, 1990; and Bernal Perez, 2001: 10). Recall from Chapter 2 that the linguistic notation allows linguists to correctly pronounce the expressions. For us to understand those pronunciations, we would need to consult a linguist or a linguistic source of the notation. You can look back to Chapter 2 regarding this topic. A comparison of Otomi dialects in Barriga Puente's work (he lists the Otomi dialects as linguistic category 49) shows that the Mezquital Valley dialect has had less influence from Spanish than most of the other Otomi dialects.

For the numbers from one to five, the only connection between terms that appears is the possibility that *goho* for four might represent twice two. On the other hand, starting at the term for six, there begins to be a pattern. Notice that the word for six, *ˀrato*, can be split into the parts *ˀra* and *to*. The part *ˀra* is similar to *ʔna* from the number one, but it is a good idea to look further down the list before concluding anything. The word for seven, *yoto*, splits into *yo* and *to*, and the word for eight, *hñato* splits into *hña* and *to*. From these examples, we start to see part of the word for two in the word for seven, part of the word for three appears in the word for eight, and part of the word for four appears in the word for nine. The word for ten no longer follows the pattern; however, the word for eleven appears to represent the construction of "ten plus one." We can now conclude that the *ʔra* in the term for six is probably "one," and that *to* means "to add to five." Hence, six is "one plus five," seven is "two plus five," eight is "three plus five," and nine is "four plus five."

Let us now look again at the expression for eleven, *ʔrœtʔ a ma- ʔra*. It is easy to deduce that this expression represents "11 = 10 + 1," and similarly, "12 = 10 + 2." It turns out that this pattern continues until twenty, when a new word, *ʔnate*, appears for that value. A continuation by ones occurs with twenty-one being expressed as 20 + 1, until thirty. The term for thirty is 20 + 10, as opposed to 3 × 10, and this implies base 20 use. When we get to 40, the term *yote* is clearly not 30 + 10. On the other hand, the prefix *yo* matches the prefix *yo* in the word *yoho* for two. It must be that *yote* is in fact 2 × 20. Later, you can see that 50 = 40 + 10 = (2 × 20) + 10, and so on. This pattern continues until 100 when the new word *nthebe* for hundreds appears. Later, the word *ʔna* followed by *mahua̧hi* appears, which corresponds to 1000. You can see some details of other values until the expression for 100,000, which is *ʔna nthebe nehua̧hi*. Observe that the prefix of *ma* from 1000 changes to *ne* for 100,000. This is because *ne* is used as a connector between units of hundreds with smaller units, while *ma* is used as a connector between units of tens with smaller units (of ones). In fact, this is explained as the rule for larger Otomi numbers in *Luces Contemporaneas* (1979: 73–74).

These constructions of Otomi numbers have been known for some time. Specific details of this counting system can be found in the references mentioned regarding Table 9.1. Nevertheless, we can ask ourselves here, "What is the counting base for the Otomi number system?" Indeed, there are values based on five (for numbers six to nine), while other values are constructed using ten and twenty. So, our conclusion is that Otomi counting does not use a single base but rather a combination of bases. In particular, the Otomi system would be called a 5-10-20 system (see Closs, 1986: 3). The Otomi system follows the counting patterns of several Mesoamerican cultural groups. For example, the Aztecs use a similar 5-10-20 system (see, e.g., Closs, 1986: 214–218). The Mayans use a 10-20 system (Closs, 1986: 292–293). Examples of counting systems of other Mesoamerican groups can be found in Barriga Puente (1998). Think back to Chapter 2 on numeration systems and the variety of ways that they can be created.

Now that we know the basic structure of Otomi numbers, our next project will be to construct some specific Otomi number expressions. This will help us understand the Otomi counting system better.

Example 9.1 Determine an Otomi expression for the value of 801.

We can write this value as $8 \times 100 + 1$. We can look at Table 9.1 and see that the word for 100 is *nthebe*. The word *ʔna* represents "one," but we want to express eight here. We can add a prefix of the word *hñato* for eight to the word for 100. To finish, let us add one by inserting *ma- ʔra*. The final expression is: *hñato nthebe ma- ʔra*.

Example 9.2 Determine an Otomi expression for the value of 270,426.

We can write this value as $(2 \times 100 + 7 \times 10) \times 1{,}000 + (4 \times 100 + 26)$. We can look at Table 9.1 and see that the word for 1,000 is *mahua̧hi*, and we have 270,000. A careful look at Table 9.1 reveals our previous observation that when the units of thousands are measured in hundreds, the prefix *ma* changes to *ne*. So, our construction will be to form the Otomi expression for 270, then align it with *nehua̧hi*. Recalling that the structure for 70 is $3 \times 20 + 10$, we express 70 with these terms. Thus, we can express 270,000 as *(yoho nthebe ne hñu ʔrate ma- ʔrœtʔa) nehua̧hi*. I have included parentheses to mimic the structure $(2 \times 100 + 7 \times 10) \times 1{,}000$. Now we put in the part of 426, connecting it with the Otomi word *ne* as an addition. Four hundreds is *goho nthebe* and we connect again with *ne* to 26, the value of 26 being *ʔnate ma- ʔrato*. The final expression is:

> 270,426 = *yoho nthebe ne hñu ʔrate ma- ʔrœtʔa nehua̧hi ne goho nthebe ne ʔnate ma- ʔrato*.

It appears that the Otomi expression is rather cumbersome because it contains a lot of words, but in many languages, expressing large values typically requires several words in order to express each part of the value. Before another example, it might be worth comparing the above expression with how it would be stated in English. It would be:

> 270,426 = *two hundred seventy thousand four hundred twenty-six*.

You can see that the expression also requires several words.

Example 9.3 Determine an Otomi expression for the value of 7,453,961.

To start, we construct 7,000,000 as 7 times one million: *yoto mahua̧hi de ga mahua̧hi*. Next, we have 453 times a thousand, which we can construct as *goho nthebe ne yote ma-ʔrœtʔa nehua̧hi* (recall how *ne* is used to connect terms of hundreds to terms of tens). Finally, we need 961: *gïto nthebe ma hñu ʔrate ma- ʔna*. Putting it all together, 7,453,961 is expressed as:

> *yoto mahua̧hi de ga mahua̧hi goho nthebe ne yote ma- ʔrœtʔa nehua̧hi gïto nthebe ne hñu ʔrate ma- ʔna*.

We can do things in reverse order, too. That is, given an Otomi expression for a number, we can construct what value it represents. Let us try a few examples of this.

Example 9.4 Determine the value of the Otomi expression *yote ma-yoto*.

By looking at Table 9.1 we see that *yote* corresponds to forty, while *yoto* is seven. Thus, this value is forty-seven.

Example 9.5 Determine the value of the Otomi expression *kïtʔa nthebe ne goho ʔrate ma- ʔrato nehua̧hi hñato nthebe ne hñu ʔrate ma-hñu.*

By looking for terms representing powers of ten, we can break the overall expression into smaller expressions. The second word, *nthebe*, indicates a term of hundred, followed five words later by *nehua̧hi*, which indicates some term of hundreds multiplied by 1,000. We can determine from Table 9.1 that *kïtʔa* is five, and *goho ʔrate ma- ʔrato* represents $4 \times 20 + 6$. This tells that the first part of the expression is $586 \times 1{,}000$, or, 586,000. Continuing to the next terms we see that *hñato* for eight multiplies 100, then is added to $3 \times 20 + 3$. We piece all of this together to conclude that the second part of the expression refers to 863. The final value described by the expression is 586,863.

Notice how we begin to think in terms of base twenty (60 as three twenties), and that clues such as the term *ne* tell us about connections with terms of hundreds.

At the beginning of this chapter I mentioned that we will find it useful to consider Aztec mathematics as a comparison to Otomi mathematics, as a way to distinguish the mathematics of each culture. Let us begin by carefully constructing some Otomi words. Then we will compare the Otomi system with the Aztec counting system. Table 9.2 shows the details of some Otomi number expressions.

Because the structure of (counting) words is a linguistic consideration, the Aztec words are described in terms of their language, Náhuatl (pronounced "NAH-wah-tl"). Table 9.3 has a list of Náhuatl numbers words, based on descriptions in Payne and Closs (1986: 214–218).

We can see that the Aztec counting system includes a factor of 15 that is not present in the Otomi system. Also, it turns out that the Aztec system employs terms of 400 and 8,000 for larger values, and this is another difference between the Aztec and Otomi systems. Now, let us look for similarities.

The main similarity between the Otomi and Aztec number systems is that the structures for the expressions for six to nine have the same pattern of values being added to five. This is worth looking into deeper.

It is generally understood that the Aztecs, a Yuto-Aztecan speaking group (the term *Yuto-Aztecan* identifies their language Náhuatl in linguistic terms), migrated to present-day Mexico City from somewhere north of that location. As indicated in the introduction, the Otomies are a very old Mesoamerican culture and were already established in central Mexico when the Aztecs

TABLE 9.2 Some Otomi Number Expressions and Structures

Number	Otomi Expression	Construction
1	*ʔna*	
2	*yoho*	
3	*hñu*	
4	*goho*	
5	*küʔa*	
6	*ʔrato*	1 + 5
7	*yoto*	2 + 5
8	*hñato*	3 + 5
9	*güto*	4 + 5
10	*ʔrætʔa*	
11	*ʔrætʔa ma- ʔra*	10 + 1
12	*ʔrætʔa ma- yoho*	10 + 2
20	*ʔnate*	
39	*ʔnate ma- ʔrætʔa ma- güto*	20 + 10 + 4 + 5 = 20 + 10 + 9
63	*hñu ʔrate ma- hñu*	3 × 20 + 3
85	*goho ʔrate ma- küʔa*	4 × 20 + 5
100	*ʔna nthebe*	1 × 100
1000	*ʔna mahua̧hi*	1 × 1000

arrived. Therefore, it makes sense to examine maps of linguistic groups. In Valiñas Coalla (2000), there are several tables and maps of 16th-century linguistic information regarding number words of Northern and Southern Yuto-Aztecan. The geographical area he studies includes the present-day southwestern United States and northwestern Mexico. One such group from the northern part of this geographic area is the Opata ("oh-PAH-tah"). You may recall a reference made to the Opata in Chapter 2 on numeration. The following Opata number expressions are taken from Valiñas Coalla (2000). The expressions for one to ten are: 1, *seni*; 2, *gode*; 3, *vaide*; 4, *nago*; 5, *marizi*; 6, *bussani*; 7, *seni-bussani*; 8, *go nago*; 9, *kimakoi*; and 10, *makoi*. Notice that the word for six has no connection with the word for five. The expression for seven is: "7 = 1 + 6." For eight, we have: "8 = 2 × 4," and nine is "9 = 10 – 1." A look at the other Yuto-Aztecan dialects from the northern part of the region (roughly the part that would be in the present-day southwest of the United States), shows that most such dialects follow a pattern similar to the Opata. Meanwhile, checking the dialects that are further south, we see that most of those Yuto-Aztecan dialects have the same structures for the numbers six through nine as do cultural groups from central (present-day) Mexico, including the Otomies. Several cultural groups of Yuto-Aztecan speakers migrated

TABLE 9.3 Náhuatl Number Words

Number	Náhuatl Phrase	Construction
1	*ce*	
2	*ome*	
3	*yei*	
4	*nahui*	
5	*macuilli*	
6	*chicuace*	1 + 5
7	*chicome*	2 + 5
8	*chicuei*	3 + 5
9	*chiconahui*	4 + 5
10	*matlactli*	
11	*matlactli once*	10 + 1
12	*matlactli omome*	10 + 2
13	*matlactli omei*	10 + 3
15	*caxtolli*	
16	*caxtolli once*	15 + 1
17	*caxtolli omome*	15 + 2
20	*cempoalli*	
30	*cempoalli ommatlactli*	20 + 10
37	*cempoalli oncaxtolli omome*	20 + 15 + 2 = 20 + 17
60	*eipoalli*	3 × 20
100	*macuilpoalli*	5 × 20
300	*caxtolli onnauhpoalli*	(15 + 4) × 20
400	*tzontli*	1 × 400

toward central Mexico over the years before the rise of the Aztecs (Smith, 2003: 37). It appears that groups that migrated closer to central Mexico adopted number expressions similar to those of groups such as the Otomies. Based on these observations, we can conclude that the structures of the Aztec number expressions for six to nine represent an influence by existing cultural groups of central Mexico, including the Otomies. This conclusion is supported by Doris Bartholomew (see Bartholomew, 2000), who concludes that the Aztec expressions for six to nine were directly influenced by the Otomies. The difference between her analysis and ours is that hers is based on anthropological considerations, while ours is based on considering mathematical structures as part of *cultural mathematics*. Now it is time to move on to the next topic, artwork.

OTOMI ARTWORK

The Otomies have traditionally been highly regarded for their skills at textile arts, and I would like to examine the mathematics of Otomi weaving and embroidery. Some general reminders of our observations in Chapter 5 about weaving and embroidery are worthwhile. I will tend to refer to Otomi textile artists as female because most are women, though there are male weavers and embroiderers. Now is a good moment to be cognizant of the fact that textile art is often considered to be a leisure activity of women and not to be part of mathematics. However, we already know that creating traditional textile art mathematics is indeed taking place and that creating art in a traditional manner is not typically a leisure activity.

First, to become a traditional Otomi weaver is a life-long process. A common situation is one in which there are several generations involved in creating woven or embroidered projects. That is, there is often a gathering that includes girls, mothers, aunts, grandmothers, and so on. Second, Otomi weavers and embroiderers must keep track of many counts of threads and must make precise measurements. They must know the entire design from memory. The impressive part of the weaving or embroidery process is that the artist typically is not using diagrams for the patterns, nor is that person using a ruler to measure distances. A third consideration is that the products finished by the artists (as is the case in many traditional contexts) often have important cultural significance. Finally, it is important to mention that many kinds of weaving and embroidery designs are very time-consuming to make. It is not unusual for traditional textile artists to take several weeks to make a work of art.

We start with weaving, a textile art that is common in Otomi culture. Weaving is a practice that dates back to pre-Columbian times. As I alluded to above, the mathematical aspects of weaving involve a lot of thread counting and keeping track of those thread counts. Conversations I had with traditional Otomi weavers in Tolimán (in the Mexican state of Querétero) in 2001 revealed that in the past Otomi weavers kept track of all thread counts by memory. Jacques Soustelle (1937) published one of the most complete anthropological studies of Otomi culture, and in it he describes his observations of traditional Otomi weavers and embroiderers. We can use cultural mathematics to observe the thread counting involved in Otomi textile art by reading what Soustelle says about Otomi artists he observed: "*La base es la cuenta de hilos.... La memoria de los tejedores Otomíes es tan perfecta, sus cuentas de hilos tan exactas, que ... queda perfectamente simétrico*" (p. 94). This translates to: "[The art of weaving] is based on thread counts.... The memory of the Otomi weavers is so perfect, their thread counts are so exact, that ... [the final product] ends up perfectly symmetric." The description by Soustelle verifies to us that what the Otomi artists are doing does in fact require mathematical processes of organizing information (via the thread counts) and knowledge of concepts of symmetry.

Embroidery is a related textile art for which Otomi artists also typically have substantial skill. In the case of embroidery there are usually complex patterns, often including several figures that have properties of symmetry. So, let us look at symmetric properties of some Otomi works of art.

Figure 9.1 An Otomi cloth.

Let us look at some specific examples of Otomi embroidery. Figure 9.1 shows a cloth from an Otomi market in Ixmiquilpan, in the Mexican state of Hidalgo. If we look at the pattern of the bird images that appear in the figure, we can see that they are symmetric about a vertical line that could be drawn down the middle of the figure. Also, they are horizontally symmetric, if we draw an imaginary horizontal line through the middle of the figure. Recall these kinds of observations from our earlier discussions of art and symmetry.

In the case of the Otomies, we can concretely deduce that the person who embroidered a product such as that in Figure 9.1 does really understand the concepts of vertical and horizontal symmetry. I quote Soustelle again: "*[S]e hace generalmente el motivo de manera que quede situado justo en medio del sarape, . . . , es necesario . . . hacerlo en dos etapas, una mitad sobre el borde derecho de la banda, la otra sobre el izquierdo*" (p. 94). This translates to: "In general the pattern is made in such a way as to be situated in the middle of the shawl, . . . , and it is necessary . . . to make it in two stages, half on the right hand border and the other half on the left side." Thus, the artist creates half of the figure, then completes the other half. To make the figures symmetric, *the artist must create the second half of the figure in the proper orientation*. We saw in Chapter 5 that this means the artist must understand the symmetry properties.

Figure 9.2 Otomi pattern from a shoulder cape. (Courtesy of *Mexican Indian Folk Designs, 252 Motifs from Textiles*, Irmgard Weitlaner-Johnson, Dover Publications, 1993, plate 64, p. 24.)

As for measurements, a close look at Otomi textile work shows very precise spacing, such as the sizes of figures and the spaces between figures. Nevertheless, the artists create their works without the use of diagrams or rulers. The precise measurements are made from memory and from much practice in the embroidery process. Figure 9.2 shows another Otomi textile example. I leave it to the reader to decide which of the symmetry properties described in Chapter 5 are possessed by these examples. Remember to include alternating colors in your considerations.

Let us look an example of Otomi art considered as patterns on a strip, as we did in Chapter 5.

Example 9.6 Determine to which of the seven strip pattern groups the design in Figure 9.3 belongs.

First, we notice that colors are alternating in this design. Meanwhile, the design has horizontal, vertical, and 180° rotational symmetry. This means, according to Table 5.1 of Chapter 5, that the design belongs to type 7.

There are more examples for you to consider in the exercises. Our next topic is that of Otomi number symbols.

Figure 9.3 Otomi pattern from a shoulder cape. (Courtesy of *Mexican Indian Folk Designs, 252 Motifs from Textiles*, Irmgard Weitlaner-Johnson, Dover Publications, 1993, plate 66, p. 25.)

OTOMI USE OF NUMBER SYMBOLS

We want to know what kinds of number symbols the Otomies would have used. It turns out that we can answer this by learning some things about the Aztecs.

Part of the long and complicated history of interaction between the Aztecs and Otomies involves the fact that much of Otomi territory was eventually controlled by the Aztecs, which led to ongoing conflict between the two cultures. The Aztecs collected tributes from groups they controlled; descriptions of such tributes appear in various codices relevant to Aztec history. Several such codices contain mathematical information, and we will see that this information tells us something about Otomi mathematics.

First, a description of the mathematical symbols that appear in such codices is shown in Figure 9.4. It is generally understood that the Aztecs created these symbols. The symbols appear most often in the context of tributes that were given to the Aztecs from people in territory they controlled and, later, in the context of tributes given to local Spanish officials.

Figure 9.5 shows a folio (page) from the *Mendoza Codex* in which some mathematical symbols appear. This codex was written by a person of Aztec culture, shortly after the Spanish conquest of Mexico (see Ross, 1978: 11). Several folios of the codex describe tributes the Aztecs collected from towns and locations they controlled. There is also Spanish handwriting in several places, giving explanations for the images. Observe the symbols for 20 and 400 in several places. Along the left side of the folio there are ten drawings with writing above each one. These drawings are glyphs that identify which towns and locations had delivered tributes that are described on that folio. The third

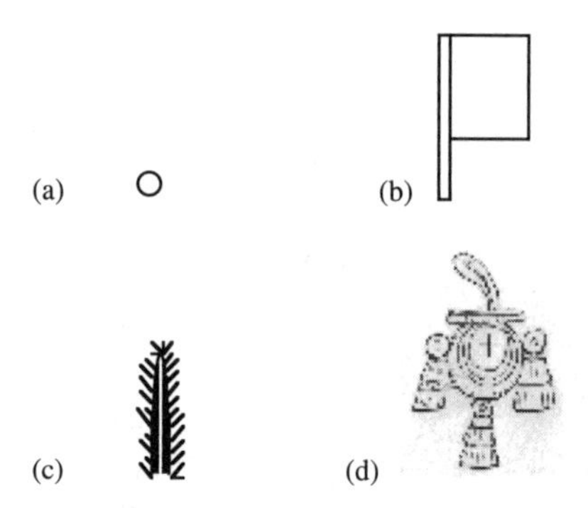

Figure 9.4 Mathematical symbols from Mesoamerican codices. (a) The symbol for 1 (units); (b) the symbol for 20; (c) the symbol for 400; and (d) the symbol for 8000. (Adapted by the author from Ross (1978: 42–56).)

and fourth glyphs from the bottom are those of Ixmiquilpan and Mixquiahuala, respectively. These two locations are in the Mezquital Valley and have been considered part of Otomi territory since before the *Mendoza Codex* was written. Thus, in this folio some of the tributes were delivered by Otomies to the Aztecs. We can reasonably conclude that the Otomies involved with those tributes must have understood the mathematical symbols that were used to represent the quantities. We could even go a bit further by realizing that many people who managed tribute payments to the Aztecs must have understood the numerical symbols. Because the Aztecs controlled a large territory that included people from several cultures (such as the Mazahua and Matlanzinca), we could consider the mathematical symbols of Figure 9.4 as representing a mathematical use for a large part of present-day central Mexico. It would be more descriptive for us to refer the symbols as *Aztec-Mesoamerican* number symbols.

Below I will describe another codex that we can use as a comparison, but first, some more details about the folio from the *Mendoza Codex* are in order. Notice the images near the top of the folio. They are square-looking images with a symbol of 400 above each one. In fact, these images represent textiles that were given to the Aztecs as part of the tributes they collected. According to Ross (1978: 37), the icons with colored patterns in them represent such things as mantles and tunics that have some pattern to them. Ross explains that such textiles were "richly embroidered" (p. 46). We can interpret "richly embroidered" to mean that the textiles contain several and/or complicated patterns. As we know from the discussion of Otomi art, this means the textiles probably included properties of various kinds of symmetry, and took a long time to make. The blank icons represent larger mantles without special patterns. There are a total of $6 \times 400 = 2{,}400$ total textiles that were delivered as tributes by the ten locations to the left. The images below the textiles represent war suits (also called war dresses) and shields. Ross tells us that tributes of

Figure 9.5 A folio from the *Mendoza Codex*. To the left are the glyphs for Mixquiahuala and Ixmiquilpan. (Courtesy of the Bodleian Library, Oxford, UK.)

mantles were collected every six months and the war suits every year (1978: 37). Two such images have the symbol for 20 connected to them, so more than 40 such suits were delivered to the Aztecs. From the previous section we know that creating these textile products is a complicated process that includes mathematical concepts and thinking. Also, making 2,420 mantles, tunics, war dresses, and shields represents a lot of work by many people! The fact that the Aztecs collected them shows that they had an appreciation of the work involved in making them, and probably held the textile artists in high regard.

Figure 9.6 shows a folio from a fragmented codex known as the *Mixquiahuala Codex*. The most recent and complete description of the *Mixquiahuala Codex* that I know of is Hermann Lejarazu (2001).

Hermann Lejarazu explains (2001: 96) that the part of the *Mixquiahuala Codex* that appears in Figure 9.6 was written in about 1571. Above each

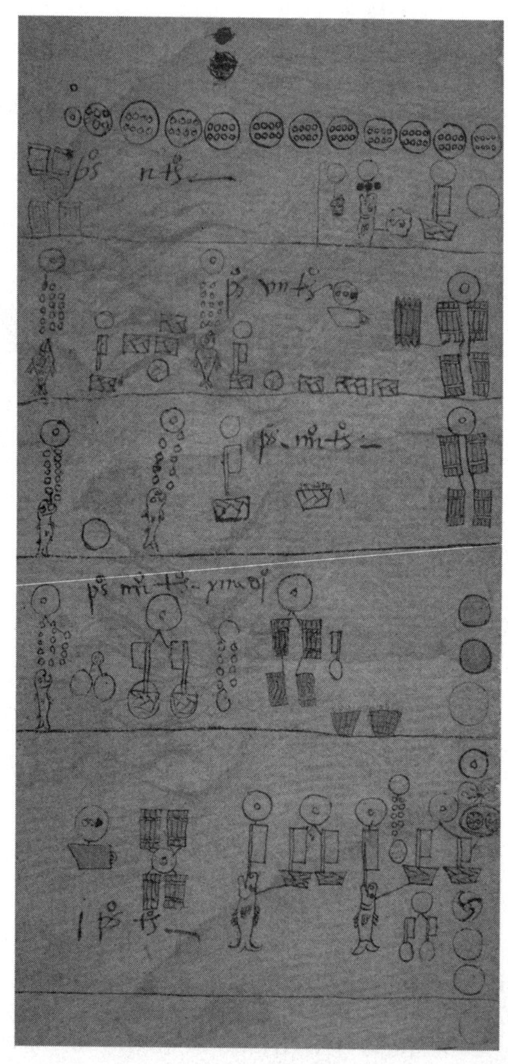

Figure 9.6 Folio from the *Códice de Tributos Mizquiahuala.* (Reproduced with the permission of the Instituto Nacional de Antropología e Historia (INAH), Mexico City.)

drawing we can see either some small circles, one or more flags, or a larger circle that may have a small circle drawn inside it. The larger circles with (in some cases without) smaller circles within them represent the amount in *tomines* (the local currency used at the time) that the local Spanish authority paid for the quantities of goods delivered. The figures immediately above each drawing are represented by either the symbol for one or flags for twenty. As an example, in the last row of figures (from left to right) you can see that there is a teapot-like image followed by notation for four bundles of firewood, 20 fish, 40 baskets of tortillas, another 20 fish, and so forth. The symbols at the end of that line, shown as three circles in the bottom right corner, describe an approximate date of delivery in the form of a representation of phases of the moon. What distinguishes this fragment from the folio of the *Mendoza Codex* of Figure 9.5 is that the *Mixquiahuala Codex* represents an Otomi-Spanish communication. We can see that the pre-Hispanic symbols were still in use at time of the writing of this codex. This gives us more confidence to conclude that the mathematical symbols of Figure 9.4 were known to and understood by people of Otomi culture. There are more examples of the use of these mathematical symbols by Otomies, such as in the *Osuna Codex* (see Valle, 2001).

THE OTOMI CALENDAR

The Otomies used a calendar system that consisted of a 365-day solar calendar and a 260-day ritual calendar. This structure is similar to many Mesoamerican calendars (see Broda de Casas, 1969), and we saw a similar structure in the Mayan calendar. The Mayans are also a Mesoamerican culture. First let us learn the basic structure of the Otomi calendar. After that, we will consider the mathematical aspects of the leap year, and the implications of this question to interpreting the mathematics of the Otomi.

The primary source for information on the Otomi calendar comes from the *Huichapan* (pronounced "wee-CHAH-pahn") *Codex*. The *Huichapan Codex* was probably written sometime shortly after 1632, according to Caso (1967: 211). It is a post-Conquest document that is written in Otomi with Spanish translations written near the Otomi text. Many anthropologists have studied it and have agreed that it was originally written by a person of Otomi culture. It was worthwhile to discuss the authenticity of the *Huichapan Codex* because, as you may recall from the Introduction to Part II, we want accurate descriptions.

The structure of the Otomi solar calendar is that of eighteen cycles of twenty days each for a total of 360 days, followed by five extra days added at the end, for a total of 365. The ritual calendar is calculated as $13 \times 20 = 260$ days. The Otomi expression for a given year appears as a drawing that consists of one to thirteen small circles placed near a glyph that represents one of twenty day names. Figure 9.7 shows two such dates from the *Huichapan Codex*.

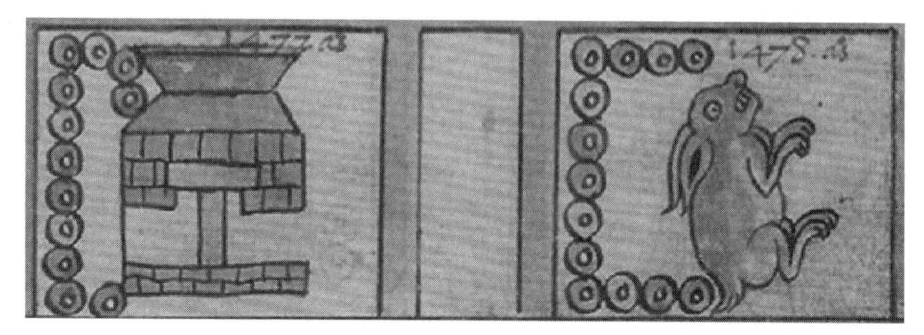

Figure 9.7 Two Otomi calendar year dates from the *Huichapan Codex*. (Reproduced with the permission of the Instituto Nacional de Antropología e Historia (INAH), Mexico City.)

From left to right, the two dates shown above can be translated to English to mean: 12 *House* and 13 *Rabbit*. We will see shortly that the explanation for using numerical values of one to thirteen and twenty day names is that by counting cycles of 365 days in this manner, it is easy to predict when the 260-day ritual calendar and 365-day solar calendar coincide. That coincidence occurs when the least common multiple (*lcm*) of both calendar cycles agree, which is:

$$\begin{aligned} lcm(365, 260) &= lcm(73 \times 5, 13 \times 20) = 13 \times 73 \times 20 = 18{,}980 \text{ days} \\ &= 365 \times 52 = 260 \times 73. \end{aligned}$$

Thus, every 52 solar years the two calendars coincide, as we saw previously with the Mayan calendar in Chapter 8.

Now let us look at the description of the Otomi years. Table 9.4 shows the list of the twenty day names used by the Otomies in their calendar. I have included the Otomi terms of the day names that appeared in the original *Huichapan Codex*. It should be noted that the person who wrote the codex was attempting to align the Otomi calendar with the European calendar of the 1600s, the Aztec calendar, and the signs of the zodiac. As such, there are some minor variations in descriptions of the day names because it is not clear whether those names represent original Otomi terms, or modified terms that were intended to make the three calendars and zodiac signs be in agreement. For more details, you may look at Soustelle (1937) and Caso (1967). I have used Soustelle's list of Otomi day names here. Also, the author of the *Huichapan Codex* attached a prefix of either "An" or "Am" to all of the Otomi terms, presumably to signify the singular grammatical case. It is not clear if the Otomies would have used these prefixes when speaking of the Otomi day names, and both Soustelle and Caso include lists of more typical Otomi words (i.e., constructed without the "an" or "am" prefixes) for the day names.

We do not have concrete examples of how an Otomi day would be described, as we had with the Mayan calendar; however, it is known that the Otomi solar calendar had the same structure of $18 \times 20 + 5 = 365$, similar to the Mayan solar cycle, and similar to several other Mesoamerican solar calendars. On the

TABLE 9.4 The Twenty Day Names of the Otomi Calendar (translated from Soustelle, 1937: 519)

Day Name	Name in Otomi
Knife	*Antoqhuay*
Wind	*Amadăhi*
House	*Anegû*
Lizard	*Anbotăga*
Snake	*Ancquĕyă*
Death, dead person	*Anyăyăy âtu*
Deer	*Anphanixantoehoe*
Rabbit	*Anqhua*
Water	*Andehe*
Dog	*Anyoh*
Monkey	*Amatzepă*
Yellow herb, grass	*Anchăxttey*
Cane (sugar)	*Anxithi*
Wild animal, ocelot	*Anhmatzhăni*
Eagle	*Angaxeni*
Turkey	*Anthecha*
Earth movement	*Anquitzhĕy*
Flint	*Aneyaxi*
Rain	*Anyeh*
Flower	*Andoeni*

TABLE 9.5 The Four Year Names (and Abbreviations) of the Otomi Calendar

Number	Otomi	Year Name	Huichapan Codex
1	*Ngû*	House	*Anegû*
2	*K'wa*	Rabbit	*Anqhua*
3	*Šit'i.*	Cane	*Anxithi*
4	*Doyaši*	Flint	*Aneyaxi*

other hand, the original author of the *Huichapan Codex* describes several years using the Otomi calendar system, so we have what is probably an accurate description of how to denote Otomi years. It should be noted here that the structure of Otomi years we are about to see is similar to the structure of denoting years in some other Mesoamerican calendars, most significantly, the Aztec calendar.

The basic structure of Otomi years is a number followed by one of four day names from the twenty day names given in Table 9.4. In English the four day names translate to: *House, Rabbit, Cane*, and *Flint*. Because these four names are the only ones used to denote Otomi years, let us organize them. Table 9.5 shows the four terms in root Otomi words (i.e., obtained by deleting the "An" prefix), the names in English, and the original Otomi terms from the *Huichapan Codex*. See Soustelle (1937: 519) for more details on the Otomi vocabulary.

I have numbered the year names from one to four, for numerical calculations coming up shortly. To describe Otomi years, I will write a number between one and thirteen followed by one of the four Otomi year names. This is how the Otomi years are stated in the *Huichapan Codex*, and it is a typical way to describe similar Mesoamerican calendars. I have chosen not to use the Otomi words that have the "An" prefix, so as to have the year names be more easily distinguishable.

We are now ready to look at how to denote Otomi years. We can arbitrarily start with our first year denoted by 1 *Ngû*. Then the year names proceed as follows:

> 1 *Ngû*, 2 *K'wa*, 3 *Šit'i*, 4 *Doyaši*, 5 *Ngû*, 6 *K'wa*, 7 *Šit'i*, 8 *Doyaši*, 9 *Ngû*, 10 *K'wa*, 11 *Šit'i*, 12 *Doyaši*, 13 *Ngû*, 1 *K'wa*, 2 *Šit'i*, 3 *Doyaši*.

You can see from the above description that after 13 the numbering starts over and that the structure is similar to the *Tzolkin* cycles of the Mayan calendar. For calculations, we can let i denote the number value and j denote the number corresponding to the year name. Thus, the values for i and j are $i = 1, 2, 3, \ldots, 13$, and $j = 1, 2, 3, 4$.

Keep in mind that this notation is a Western interpretation, and the Otomies did not calculate dates using this kind of mathematical expression.

We can create some formulas for determining an Otomi year that occurs M years before or after a given Otomi year, by observing that the index i repeats after 13, while the index j repeats after four. Thus, for M years before or after an Otomi year given by ${}_iO_j$, the new indices i_M and j_M will be determined by:

$$i_M = (i + M)(\text{mod } 13), \text{and } j_M = (j + M)(\text{mod } 4).$$

These are the same kind of equations as for the Mayan calendar calculations of Chapter 8, with the exception that I have used M here in place of N. It is worthwhile to refresh our memories about these kinds of problems, which were dealt with in Chapter 8. Recall that our Western method of solving $x = A(\text{mod } n)$ is to create an equation based on whether A is positive or negative:

$$x = A - n \cdot k, \text{for } A > 0,$$

$$x = n \cdot k - A, \text{for } A < 0,$$

where we seek integers k whose multiples of n will give us a value of x that is in the range of $1 \leq x \leq n$.

Let us go through a few examples of the calculations involved with Otomi years. If you recall the material from the Mayan calendars, you will notice that the process is similar.

Example 9.7 Determine the Otomi year that would be 14 years after 2 *Ngû*.

Recall from Chapter 8 on calendars that we can determine the information by putting the known values into the two general equations above. With $M = 14$, $i = 2$, and $j = 1$, we can write:

$$i_M = (2+14)(\text{mod } 13) = 16(\text{mod } 13),$$

and

$$j_M = (1+14)(\text{mod } 4) = 15(\text{mod } 4).$$

We first solve $i_M = 16(\text{mod } 13)$, noting the obvious fact that $16 > 0$. Thus, from above, we set up the equation

$$i_M = 16 - 3 \cdot k.$$

If we try $k = 1$, we get: $i_M = 16 - 13.1 = 16 - 13 = 3$.

Next, we solve for j_M in $j_M = 15(\text{mod } 4)$. This is equivalent to $j_M = 15 - 4 \cdot k$. With the same logic as for the case of i_M, choosing $k = 3$, we obtain $j_M = 15 - 12 = 3$. Now, the third year name is *Šit'i*, so our final expression is the Otomi year 3 *Šit'i*.

Example 9.8 Determine the Otomi year that would be 42 years before 7 *Šit'i*.

This time we are looking into the past, and we wish to determine i_M and j_M. Now we opt for the second type of equation, that is, for the case of a value $A < 0$: $x = n \cdot k - A$. Our problem is to solve for i_M: $i_M = (7 - 42)(\text{mod } 13) = (-35)(\text{mod } 13)$, so we solve the equation

$$i_M = 13 \cdot k - 35.$$

Observe that if $k = 3$, we have $i_M = 13 \cdot 3 - 35 = 4$.

Next, for j_M, notice that *Šit'i* is the third year name, so our current value of j is 3. We solve $j_M = (3 - 42)(\text{mod } 4) = (-39)(\text{mod } 4)$. This leads to the equation

$$j_M = 4 \cdot k - 39,$$

where the choice of $k = 10$ gives us $j_M = 40 - 39 = 1$. The year name that corresponds to 1 is *Ngû*, and i_M came out to be four. This means our new Otomi year designation is 4 *Ngû*, for 42 years before 7 *Šit'i*.

Example 9.9 Suppose this year is Otomi year 3 *K'wa*. In 100 years what will be the Otomi year?

The values we have are $i = 3, j = 2$, and $M = 100$. We solve the two congruence equations:

$$i_M = (3+100)(\text{mod } 13) = 103(\text{mod } 13),$$

and

$$j_M = (2+100)(\text{mod } 4) = 102(\text{mod } 4).$$

In the first one, we write the equation $i_M = 103 - 13 \cdot k$ and look for a multiple of 13 that is still less than 103. Note that $13 \times 7 = 91$, so use $k = 7$, obtaining

$$i_M = 103 - 13 \cdot 7 = 103 - 91 = 12.$$

For j_M, we have the equation $j_M = 102 - 4 \cdot k$, and for this situation we can choose $k = 25$ to obtain $j_M = 102 - 100 = 2$. The second Otomi year name is *K'wa*, so the new Otomi year designation is 12 *K'wa*.

By now, you may be wondering why only four day names are used in the Otomi designations. In descriptions of Otomi years, such as in the *Huichapan Codex*, there are always only four day names that are ever used to describe years: *Ngû*, *K'wa*, *Šit'i*, and *Doyaši*. Why only four? The explanation is that for 365 days, we recall the structure of the Otomi calendar as $365 = 360 + 5 = 18 \times 20 + 5$. If we think about this expression carefully, it is five cycles of twenty, followed by five more days, those five extra days also being named. This means that at the end of cycling through all twenty day names, we have exactly five more names that are used to denote the days of a 365-day year. During those five days, we would advance five names down the list of twenty day names. Because $20/5 = 4$, we always return to one of only four choices for the name of the day at which a new year begins. Observe that not only is it easier to have to remember only four names for any given year, but this way of using four day names to denote years and up to thirteen numerical symbols also means that the solar and ritual calendars will coincide when the first pattern repeats; that is, when the thirteen number symbols have cycled four times for a total of $13 \times 4 = 52$ solar years.

We have resolved the question of why there are only four Otomi day names used to designate any given year in their calendar, but that explanation generates another question: If the actual length of a solar year is approximately 365.2422 days, then what happens to the Otomi calendar after many years? Indeed, this is the question of leap year in the Otomi (or any other) calendar. Because the writing of the *Huichapan Codex* and the declaration in Europe of the Gregorian calendar (by Pope Gregory VIII in 1582) occurred around the same time, there was probably a strong interest by early Spanish in the Americas as to whether calendars of indigenous groups had made adjustments to their calendars to account for what is referred to in Western calendars as leap year (where an extra day is added every four years, with further minor adjustments).

In the case of the Otomi calendar, the author of the *Huichapan Codex* attempts to explain this in Otomi on folio (page) 13 (see Lastra and Bartholomew, 2001: 43, for more details): "*N˜uccãdaandaqhueya. Edettatemahiãntitemaquüttamapa. ccclxv yquüttzi edato oras. ãnãbeattegui. Nãh˜u oras oras. Emãh˜equi e e oras.*" This statement is followed by written comments in Spanish: "*Cada año [consta de] dieciocho veintanas más cinco días [= 365] ccclxv . . . se añaden seis horas [del] reloj . . . , se llaman horas iguales horas.*" This translates into English as an explanation of a calculation of leap year by adding six hours to each year, the total over four years giving $6 \times 4 = 24$ hours. This would seem to imply a calculation of leap year in the Otomi calendar. Although there seems to be some logic in this description, from a cultural-mathematical point of view, it is doubtful that this calculation occurred. How so? The first problem is that there is no evidence that the Otomi divided their days into smaller units ("hours"). Though such a division into smaller time units is certainly possible (e.g., by observing shadows on a fixed object), the Otomies would not have made a calculation of $6 \times 4 = 24$ in this context. Can you think of why? Well, we have seen that the Otomies use a 5-10-20 number system, so using a value of 24 would not be a natural choice. If the Otomies had a version of hours to their days, it would most likely be based on 5 units per day, 10 units per day, or 20 units per day, but not 24 units per day.

The fact that only four names are used to describe years also tells us something about the possibility of adding a day to calendar after every four years. Adding one more day after four years would break the pattern of using only four day names to describe years. For example, if at the end of a year that is designated as, say, 8 *Doyaši* there would be a leap year calculation, the following year would then have a number value of 9 but a day name corresponding to an Otomi word for *Rain* because of the addition of one day (*Doyaši* is the word for *Flint* and the day name that comes after *Flint* is *Rain*). Then all the day names for the year following that leap year addition would be moved down one name on the twenty-day name list. However, as far as we know the Otomies always used the same four day names in their descriptions of years.

Thus, our conclusion about the $6 \times 4 = 24$ statement that appears in the *Huichapan Codex* is that the Otomies did not measure leap year in a way similar to how it is done in the Western calendar. In fact, we can observe a Western view of mathematics here: An assumption had been made during the time the *Huichapan Codex* was written, that all cultures must eventually make an adjustment in their calendars that would be equivalent to the Western leap year.

Now, we have addressed the question of whether the Otomies calculated leap year in their calendar—they did not—but this generates yet another question: Did the Otomies account for a solar year that is not exactly 365 days in some other way? It turns out that it is widely known that many Mesoamerican cultures were very knowledgeable of the cycles of the sun, the moon, and Venus, as well as other astronomical phenomena such as equinoxes and solstices. Moreover, the Otomi calendar was very closely connected to agricultural cycles, as described, for example, in Albores (2006). Hence, in order to maintain their calendar as an accurate tool for agriculture, the Otomies must have adjusted it in some way, from time to time. How were the adjustments made? We do not know at this time. This question is discussed in Broda de Casas (1969), and Beatriz Albores (Albores, 2006) describes a possibility that some Mesoamerican cultures related to the Otomies may have used the cycle of Venus to adjust their solar calendar.

OTOMI DIVINATION

There is some information from anthropologists regarding divination in Otomi culture, such as in Sandstrom (1981) and Dow (1996). Most of what has been observed is that there does not appear to be a mathematical algorithm involved as we saw for the Caroline Islands or *sikidy* divination processes in Chapter 6. Nevertheless, there are some mathematical observations that can be made. For example, many of the paper figures made by Otomi diviners include the number four in several places, indicating a ritual importance of the number four in the Otomi divination process (see, e.g., Sandstrom, 1981: 90, fig. 18). Moreover, we can make the same observation about Otomi divination as we did about the Caroline Islands diviners: The Otomi diviners use problem-solving skills, must have expertise in several aspects of typical life (e.g., human relations), and use a certain amount of psychology and/or social work skills in their process. These skills are analytical, even if explicit mathematical formulas or procedures are not being invoked in the process.

OTHER TOPICS

We have not discussed some of the other cultural mathematics topics from Part I of the book, with respect to the Otomies. Such topics would include kinship relations and games. This is because, as of this writing, careful mathematical studies of these aspects of Otomi culture have not been done.

We have looked at many questions regarding Otomi mathematics. Now it is your turn to think about some other questions, as provided in the exercises below.

EXERCISES

Short Answer

9.1 The mathematics of several Mesoamerican cultures (such as the Aztecs, the Mayans, and the Otomies) have similarities. Nevertheless, what is it about the Otomi language that might give us new and/or interesting information about Mesoamerican mathematics?

9.2 The author of a book on Otomi counting suggests that, although Otomi expressions could be used to describe large numbers, it is easier to use Spanish constructions. What conclusions can you draw from this statement?

9.3 In the Otomi number expressions there is an influence of base 10 counting (which existed before the arrival of the Spanish). Make a conjecture as to the origin of the use of base 10 in the Otomi numeration system.

9.4 **(a)** Why do you think the author of the *Huichapan Codex* included the statement of six hours added to the last day of each year as a way to include leap year in the Otomi calendar?

(b) Which of the Western views of mathematics is expressed in the statement about six hours added to the last day of the year? Explain.

9.5 No explicit mention of a ritual calendar of 260 days (similar to the Mayan and to many other Mesoamerican cultures) is made in the *Huichapan Codex*, but the Otomies undoubtedly had such a calendar. Based on what we have discussed in this chapter, how do we know the Otomies must have had a ritual calendar?

9.6 **(a)** Is the Otomi calendar based on physical, natural, or cultural cycles, or combinations of these? Explain.

(b) Explain why, in the numbering of years of the 365-day Otomi calendar, only four day names appear.

9.7 What does the numerical information (i.e., the quantities and types of products) from the *Mendoza Codex*, shown in Figure 9.5, say about the importance of the mathematical knowledge of Otomi women?

9.8 It turns out that the Aztec calendar is very similar to the Otomi calendar, and some anthropologists have raised the question of which culture might have developed the first version of the type of calendar we have studied in this chapter.

(a) Give one reason that would support a claim that the Otomies knew of or developed their calendar and that the Aztecs later adopted it.

(b) Give one reason that would support a claim that the Aztecs knew of or developed their calendar and that the Otomies later adopted it.

Calculations

For Exercises 9.9—9.12, determine an Otomi expression for the given value. Be sure to explain your reasoning by describing the different parts of the expression, as we did in Examples 9.1–9.3.

9.9 Determine an Otomi expression for the value 42.

9.10 Determine an Otomi expression for the value 635.

9.11 Determine an Otomi expression for the value 283,000.

9.12 Determine an Otomi expression for the value 5,689,402.

For Exercises 9.13–9.16, determine the numerical value of the Otomi expression. Be sure to explain your reasoning by writing out the powers of ten and the values multiplying them, as we did in Examples 9.4 and 9 5.

9.13 The Otomi expression *ʔrœtʔa ma- yoho.*

9.14 The Otomi expression *yote ma- ʔrœtʔa ma- hñu.*

9.15 The Otomi expression *hñato mahua̧hi ne-yoto nthebe ne-kïtʔa ʔrate ma- hñato.*

9.16 The Otomi expression *kïtʔa mahua̧hi de ga mahua̧hi ʔrato nthebe ne yote ma- ʔrœtʔa nehua̧hi hñato nthebe ne yoho ʔrate ma- gïto.*

For Exercises 9.17 and 9.18, write the Aztec-Mesoamerican symbols that represent the value described. Work with the symbols; do not convert to decimal notation.

9.17 Suppose that some tributes have values as follows:

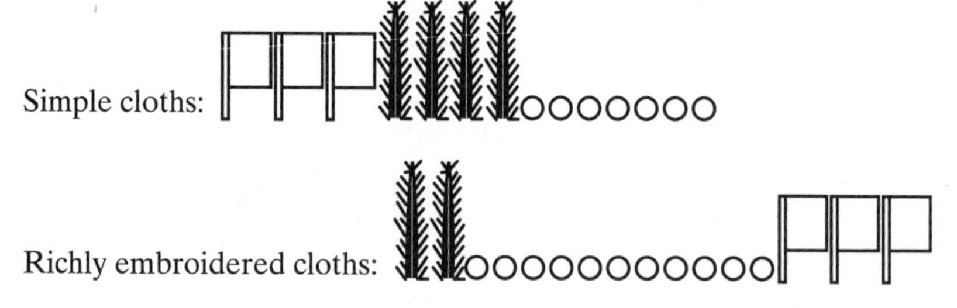

Determine and express, as Aztec-Mesoamerican symbols, the total number of cloths that were given as tributes.

9.18 **(a)** Determine a representation of the symbol in terms of the symbol .

(b) Determine a representation of the symbol  in terms of the symbol 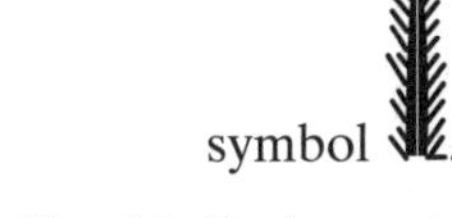.

9.19 **(a)** Is the system of Aztec-Mesoamerican symbols positional or not? Explain.

(b) From your response to part (a), would a Mesoamerican culture that used the symbols be able to develop mathematics that is advanced (in the view of Western mathematics) using those symbols? Explain, including descriptions and examples of how one could express, for example, sums and differences, products and quotients, or rational and irrational numbers.

For Exercises 9.20 and 9.21, determine the symmetry properties of the figure. Explain your conclusions.

9.20 From an Otomi textile, as shown in Figure 9.8.

9.21 From an Otomi textile, as shown in Figure 9.9.

For Exercises 9.22 and 9.23, determine to which of the seven symmetry types the strip pattern of the figure belongs. Explain your conclusions.

9.22 An Otomi-Mazahua sash, as shown in Figure 9.10.

9.23 An Otomi-Mazahua sash, as shown in Figure 9.11.

Figure 9.8 Otomi textile. (Courtesy of *Mexican Indian Folk Designs, 252 Motifs from Textiles*, Irmgard Weitlaner-Johnson, Dover Publications, 1993, plate 36.)

Figure 9.9 Otomi textile. (Courtesy of *Mexican Indian Folk Designs, 252 Motifs from Textiles*, Irmgard Weitlaner-Johnson, Dover Publications, 1993, plate 41.)

Figure 9.10 Otomi-Mazahua sash. (Photo: David Lavender, from *Arts and Crafts of Mexico*, Chloë Sayer, Thames & Hudson, © 1990a, plate 40, p. 44.)

Figure 9.11 Otomi-Mazahua sash. (Photo: David Lavender, from *Arts and Crafts of Mexico*, Chloë Sayer, Thames & Hudson, © 1990a, plate 40, p. 44.)

For Exercises 9.24 and 9.25, determine the Otomi calendar designation for the given information, as we did in Examples 9.7–9.9 of this chapter.

9.24 If a given year is 11 *Doyaši*, what will the Otomi year designation be in 50 years?

9.25 On folio (page) 24 of the *Huichapan Codex* the Western calendar year of 1423 is given as an Otomi calendar year of 9 *Šit'i* (9 *Caña* in Spanish is written in the codex. Caña in this context means sugarcane in English). Determine the following Otomi calendar designations starting from that year:

(a) The Western year of 1519.

(b) The Western year of 2000.

(c) The Western year of 300 BCE.

Essay and Discussion

9.26 Based on your solutions to Exercises 9.20–9.23 (or whichever of these you worked out), describe how the symmetry properties of Otomi artwork reflect how the Otomies view their universe. You could, for example, compare this outlook with how another culture views its universe through an expression of symmetry in its art (see Chapter 5 on art and decoration, for example).

9.27 Write about or discuss a comparison between the Otomi number system and the number system of another culture (be sure to specify the culture and some basic information about that culture).

9.28 Investigate and write about or discuss a comparison between the Otomi and Aztec cultures, including factors such as when each culture first originated, cooperation and conflicts between these two cultures, and what became of each after the arrival of the Spanish. Use this basic information to describe a mathematical comparison of these two cultures, regarding their counting systems, calendars, and so on.

9.29 Write about or discuss a comparison of the Otomi calendar with the calendar of another culture. See Chapter 8 for examples or look at outside resources.

9.30 Another *Mixquiahuala Codex* fragment (Hermann Lejarazu, 2001) showing tributes to local Spanish has virtually no products of the types described as tributes that were collected by the Aztecs. That is, there are no cloths or other such products. Essentially, the only "product" that was used as payment to the local Spanish authority was the services of women who did housework for the Spanish person. Write about or discuss the aspects of gender and mathematics in this situation. Include factors of Western views of mathematics in your discussion.

9.31 *Research paper*: Choose a culture and investigate its mathematics as a case study such as we have done here. Keep in mind some of the resources, along with their advantages and disadvantages, that were discussed in the Introduction to Part II of this book.

Additional Comments

- The most reliable earlier references to the Otomies are Soustelle (1937) and Carrasco (1950). An excellent modern treatment of the Otomies can be found in Lastra (2006).
- The works of Sayer (1990a and b) have wonderful photographs of indigenous artwork from Mexico, including that of the Otomies.
- My sources for the Otomi calendar have been Soustelle (1937), Caso (1967), and Broda de Casas (1969), where one can find much more information than what is presented here.
- Detailed descriptions of the *Huichapan Codex* can be found in Caso (1967) and Lastra and Bartholomew (2001). The latter describe the codex in detail from an anthropological perspective that originated with work by Lawrence Ecker; Caso's description is mainly focused on the aspects of the Otomi calendar.
- As is the case for many cultures, a description of number expressions that is created by linguists or anthropologists usually has a list of only the most illustrative expressions. The motive of linguists and anthropologists is to understand the language and/or culture, but our motivation here is to include an understanding of cultural mathematics as well. To this end, I have included many more values of Otomi expressions in this chapter. As I alluded to in the discussion of Otomi number expressions, the ones I included may be slightly different from the number expressions that might be used by present-day Otomies, including the factor of various dialects of Otomi that exist.
- For the Otomi calendar description, I originally wrote the four day names in English, but then changed them to Otomi. This was to give you the reader a little more insight into the language of Otomi, and the chance to learn a few new words from another language in the process.
- One could ask questions about the Otomi calendar such as, "If this year is 10 *K'wa*, when will the Otomi year of 13 *Šit'i* occur?" Such questions are natural in the realm of calendars, but as problems to be solved using Western mathematics, they require simultaneous systems of equations from modular arithmetic. Although these problems are beyond the scope of this book, you may wish to work them out on your own.

10

TAWANTINSUYU MATH: THE INCAS

Motivational Questions: Before you begin reading this chapter, take some time to think about your responses to the following questions.

- *Who were/are the Incas?*
- *How would you determine the numeration system of the Incas?*
- *After the arrival of the Spanish in the 1500s, Inca culture changed. In what ways would changes to Inca culture affect Inca mathematics?*
- *What are some obstacles to understanding Inca mathematics that could arise from interactions between Inca culture and other cultures? Are there any positive aspects (regarding mathematical ideas) to such interactions?*
- *Suppose an archaeologist has recently found an artifact in what is known to be Inca territory. Must that artifact be from the Incas? What other possibilities are there?*
- *Suppose we want to represent numbers as knots on strings. How could we represent values such as 17, 3, 2/5, 0, –14?*
- *What does it mean to have a writing system? Must a writing system have things like an alphabet and grammar rules?*
- *Before the arrival of the Spanish to Inca territory in the 1530s, a European invasion had already taken place in that part of South America. What was that invasion?*

Introduction to Cultural Mathematics: With Case Studies in the Otomies and Incas, First Edition.
Thomas E. Gilsdorf.

INTRODUCTION

In this, the second of our case studies into a particular culture, we will look at the mathematics of the Inca culture. The same initial comments are valid here as were made in the Introduction to Part II. We will discuss some general ideas concerning Inca culture and territory, then discuss the Inca numeration system and its cultural implications, Inca art and decoration, the Inca calendar, and, finally, Inca mathematical symbols. We will encounter obstacles comparable to those we saw in our study of the Otomies, namely, difficulties in obtaining accurate information, in sifting out cultural tendencies, and in determining whether or not a mathematical object or concept is from Inca culture.

INCA LANGUAGE AND CULTURE

The first term we must clarify is "Inca," by which we refer to a collection of many groups that had a common government and language but were of distinct cultural origins. When we speak of the "Inca Empire," we refer to the territory controlled by one group of people (the original Incas) from about 1100 to 1560 CE. The first Incas started near Cuzco of present-day Peru and persistently moved on neighboring groups until they controlled an enormous territory that included most of present-day Peru, plus significant parts of Ecuador, Bolivia, Chile, Argentina, and southern Colombia (see Fig. 10.1).

In 1531 the Spaniard Francisco Pizarro landed in what is now the Pacific Coast of Ecuador, and made his way to the main part of Inca territory, taking control in 1532. There were numerous previous Spanish expeditions originating in the 1520s in present-day Mexico and moving south through what is now Panama. Through human contact European diseases such as smallpox were spread that had an impact on the Incas, even before the Spanish arrived in the 1530s. More details of the Spanish arrival in Inca territory event can be found in D'Altroy (2003: chapter 13).

The Inca territory included regions of widely varying geography: coastal desert, high rugged Andes, the Lake Titicaca intermountain area, the Pacific sea coast, and the jungle on the eastern edge of the Andes as they descend into the Amazon basin. It is worthwhile to emphasize just how varied the geography is by noting that a 200-kilometer trip by air from ocean to forest in central Peru would cross twenty of the world's thirty-four major life zones. In addition, the Andes range contains many areas that are rugged enough to be inaccessible even by horseback, hence only attainable by foot.

In terms of weather, there are numerous, unstable, and/or challenging climate and geological patterns such as earthquakes, droughts, floods and the corresponding effects of periodic occurrences of El Niño.

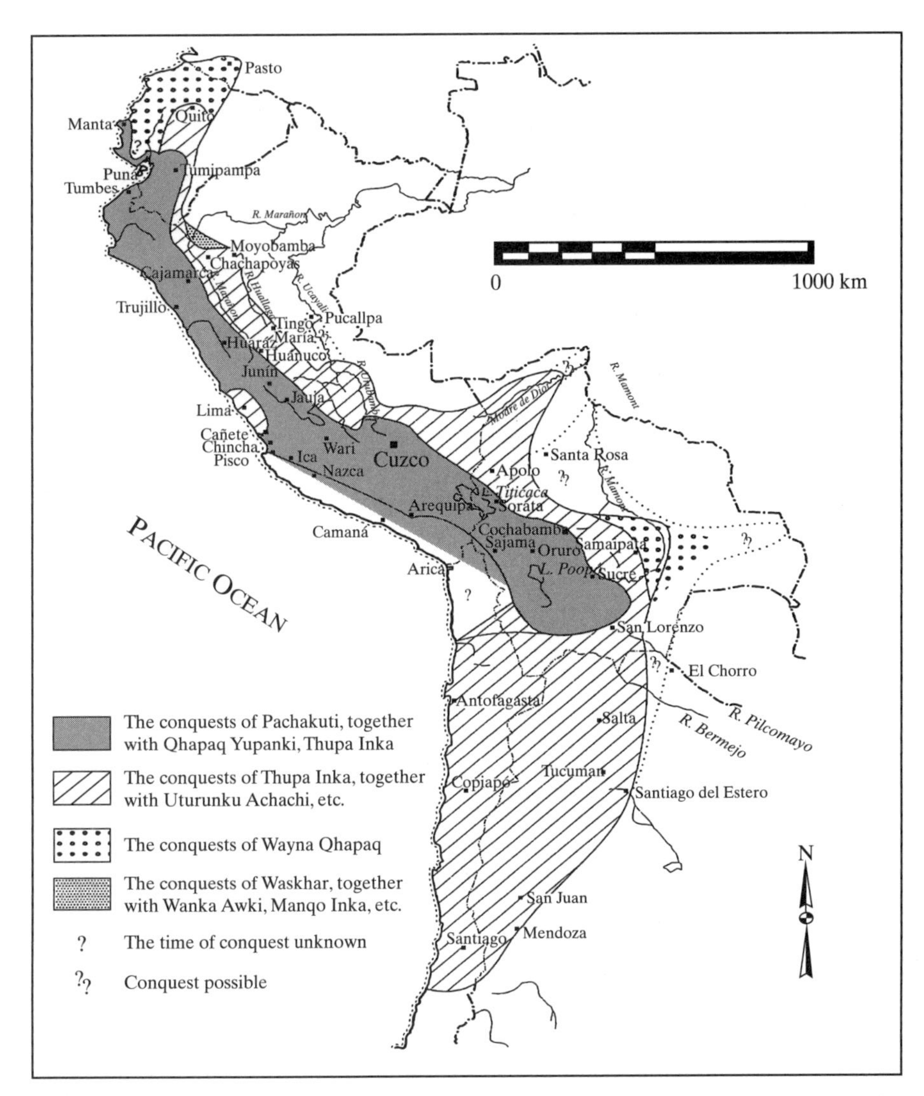

Figure 10.1 The Inca empire. (Reproduced with permission from *The Incas (Peoples of America)*, Terence D'Altroy, Oxford: Wiley-Blackwell, 2003, figure 4.1, p. 66.)

These geographical aspects indicate several opportunities for developing mathematics. For example, all successful groups in the dry regions had to have some kind of effective water control in the form of irrigation and aqueducts. Those in the high altitudes had to have some form of flexible mountain agriculture, such as terrace farming. Moreover, in many parts of the Inca territory, the groups living there had to construct bridges to cross deep canyons and difficult mountain areas. In effect, civil and agricultural engineering were crucial elements of survival. As in almost any cultural group, knowledge of astronomy was important in terms of predicting planting and harvesting seasons, anticipating weather changes, and general time keeping. Looking back to Chapter 1 and the Bishop's six categories of mathematics, we can identify such items as locating and the geometric aspects involved with it, numbers, measuring, designing, and explaining. Indeed, the construction of aqueducts, bridges, and irrigation systems involves all of these aspects.

There were many groups that we are including under the name of the Incas, and I would like to mention a few of them. The relevance is that these groups had their own mathematical concepts and practices, and as the Incas absorbed these groups into their system, they almost certainly made use of some of that mathematical knowledge.

Of the many competing cultural groups in what became Inca territory, we can first mention the *Moche* of the northern Peruvian coast, from about 100 to 700 CE. Later, in old Moche territory, a substantial group called the *Chimú* arose, forming the Chimor empire with its capital city of Chan Chan. The Moche-Chimú group was the largest Inca rival, and they inhabited the northern Peruvian coast during the Early Intermediate to Late Intermediate period (about 0 to 1470 CE). They practiced agriculture in several of the river valleys that descend from the Andes to the Pacific, including the Moche River. They were not overcome by the Incas until 1470.

Another group is the *Huari* (also *Wari*) of the central Andean highlands of Peru during the Early Intermediate and Middle Horizon periods (about 600 to 1000 CE). The Huari engaged in advanced irrigated terrace farming.

The *Aymara* inhabited the Lake Titicaca region during the Late Intermediate period (about 1000 to 1400 CE). The Aymara themselves consisted of several groups, for example, the *Colla* and *Lupaqa* groups. They had their own language group, and some dialects of Aymara are still spoken today. The Aymara also had an extensive terrace agriculture as well as domesticated llamas and alpacas.

The last of this sampling of groups that we mention is the *Nazca* of the southern Peruvian coast of the Early Intermediate period (about 400 BCE to 500 CE). Some indication of Nazca understanding of geometry and astronomy can be seen in the lines and figures they created in the desert. It is still not clear what the lines represent, but there are connections between the lines and both astronomy and the rituals of the Nazca. An important person connected with the topic of the Nazca and their astronomy and mathematics is Maria Reiche (1903–1998). Reiche was a German-born teacher who moved to Peru at a young age. By the end of her career, she had played an important role in the preservation and study of the Nazca culture and lines.

All of the above groups had some knowledge of mathematical ideas. The *quipu* that we will discuss in detail later on existed long before the rise of the Incas and was known to the Huari and Aymara. As the Incas grew to control territory, their control and/or interaction with these groups must have played into their own mathematical understanding and development. We could even go much further and say that our study of Inca mathematics is only a glimpse into the larger realm of cultural mathematics of Andean societies, most of which has yet to be carefully researched.

Tawantinsuyu, "Land of the Four Quarters," is the word the Incas used to describe their own territory, and I would like to describe some features of the Inca culture. We would like to know what Inca mathematics was like before the Spanish conquest of what was Inca territory. However, we know from the Introduction to Part II that there are obstacles regarding accurate information. Because studies of Inca culture have taken place after the arrival of the Spanish in the 1500s, information about the Incas is either substantially culturally biased, as in the case of most Spanish chroniclers, or is a study of a group that has changed significantly in nature, as is the case in studying present-day descendants of the Incas. For our purposes, two sources considered to be relatively accurate are those of *Pedro Cieza de León* and *Felipe Guamán Poma de Ayala*. Some comments about each is worthwhile.

Pedro Cieza de León was a soldier in the Spanish army and was sent to various parts of Inca territory. He recorded numerous observations of Inca culture in several insightful works, between 1547 and 1550. His writings are remarkable for at least two reasons. First is the consideration that in the mid-1500s many Europeans either had only a basic education or were illiterate, and this would have been the case for most soldiers. We can deduce that Cieza de León was more astute than a typical soldier. Second is the consideration that Cieza de León's observations show an understanding of cultural differences and a certain respect for what had been a successful culture. This kind of open-mindedness is rare for mid-1500s Spanish who mainly had intentions of conquest rather than understanding, when it came to contact with non-European cultures. An excellent source for more information about Cieza de León and his work can be found in Ascher and Ascher (1981: chapter 1). In that reference, you can also find a good list of literary resources for English or Spanish versions of his work.

Felipe Guamán Poma de Ayala was an Inca descendant, living in Spanish-controlled territory in the late 1500s and early 1600s. Between 1583 and 1615 he wrote to the Spanish king, protesting the harsh treatment of Incas by Spanish rulers in what is now Peru. Actually, the letter was 1,179 pages long, and included more than 300 drawings! Considering that everything was written by hand at that time, Guamán Poma de Ayala's work is quite impressive. Moreover, his descriptions of Inca culture are considered to be accurate. Now let us go back to discussing the Inca territory.

One may ask how the Incas could manage a large territory that covered such challenging terrain. A primary factor in their management was the Incas' impressive road and communication system. Information was conveyed to all of the four corners of Tawantinsuyu by runners who passed the information from station to station, like relay runners. Similar to the Pony Express, the difficult changes in geography of the territory made human runners the most efficient and fastest form of communication.

Another feature of the Incas worth mentioning is that as they expanded into regions of other groups, they allowed those groups a certain amount of local control. There are two advantages to this attitude. One is that the subjugated groups would not be as likely to reject Inca rule. The other, relevant to cultural mathematics, is that by allowing the local groups to retain some original culture, the Incas could use and/or improve mathematical ideas of those groups. This last comment implies that some of the mathematics of the Incas was probably used and understood by people of the groups controlled by the Incas. Thus, when we speak of Inca mathematics, we are really speaking about regional mathematics of the Andes. Compare this with Otomi mathematics as part of the general mathematical picture of Mesoamerica, from the previous chapter.

Also relevant to the development of mathematics in the Inca region is that of record keeping. As we have mentioned, the Inca empire consisted of many groups that eventually were absorbed into the Inca system. Starting with very early organized groups and extending even to the present, trade has been an important factor in the societies in the Inca region. In addition, the Inca economy included an extensive taxation system. Notice something interesting here: With their attention to detail regarding tributes (taxes), the Incas exhibit an *emphasis on number*, not unlike the emphasis on number in Western culture. We will see later that the Inca system of record keeping is a rich source of information about Inca mathematics.

INCA NUMERATION

We will look at *Quechua*, the language of the Inca. Our discussion will focus on the Cuzco dialect of Quechua. See Barriga Puente (1998: 247).

A short list of Quechua number expressions appears in Table 10.1. The expressions are accurate as of the year 1976, when the list was created (see Barriga Puente, 1998: 247, and his reference). We will make our own constructions of Quechua numbers, based on the table of information we have, but it is worth pointing out that there could be slight variations in how expressions are formed. If our intention were to create a comprehensive list of accurate number expressions, we would need to consult with an expert in Quechua regarding such slight variations. There would also be questions of how the Quechua we are looking at compares with how people speak and/or write Quechua now, were we to travel to Cuzco, as well as considerations of other dialects of Quechua. In the interest of understanding the basic ideas of Inca mathematics without getting tangled up in these kinds of details, let us agree that *our constructions are likely expressions for the value in question, with possible slight variations*. Now let us look at the structures of Inca number expressions.

TABLE 10.1 Inca (Quechua) Number Expressions

Value	Quechua Expression	Value	Quechua Expression
1	*hoq*	15	*chunka pisqa-yoq*
2	*iskay*	17	*chunka qanchis-yoq*
3	*kinsa*	19	*chunka isqon-yoq*
4	*tawa*	20	*iskay chunka*
5	*piqa*	30	*kinsa chunka*
6	*soqta*	40	*tawa chunka*
7	*qanchis*	50	*pisqa chunka*
8	*pusaq*	100	*pachak*
9	*isqon*	101	*pachak hoqniyoq*
10	*chunka*	110	*pachak chunkan*
11	*chunka hoq-ni-yoq*	1,000	*waranqa*
12	*chunka iskay-ni-yoq*	1,001	*waranqa hoqniyoq*
13	*chunka kinsa-yoq*	1976	*waranqa isqon pachak qanchis chunka soqtayoq*
14	*chunka tawa-yoq*	1,000,000	*waranqa waranqa*

The general format of number expressions is the following:

[multiplier] {nucleus} (adder).

The nucleus is always a power of ten. For the examples that follow, we will employ the above notation, namely, multipliers [], nuclei {}, and adders (). In the exercises where you will try your hand at creating Quechua expressions, I will ask you to use this format, too.

Example 10.1 *isqon pachak*: [9] {100} = 9 × 100 = 900.

Example 10.2 *qanchis chunka pisqayoq*: [7] {10} (5) = 75. Note that the term for five, *piqa*, changes to *pisqa*, and is suffixed by *yoq*, which indicates the addition to 70.

Example 10.3 24,570: [[2] {10} (4)] {1000} ([5] {100} ([7] {10})) = 24,570, so the expression for this value would be: *iskay chunka tawa-yoq waranqa pisqa pachak qanchis chunka.*

Example 10.4 347,002: [[3] {100} ([4] {10} (7))] {1000} (2), *kinsa pachak tawa chunka qanchis waranqa iskay.*

Example 10.5 7,689,543: To write the structure will require several pairs of parentheses and brackets; however, the basic building blocks are the same as in the previous examples: [7] {1,000,000} ([6] {100} ([8] {10} (9))) {1000} ([5] {100} ([4] {10} (3))). Now, we write out the Quechua terms for this structure, using Table 10.1 as necessary: *qanchis waranqa waranqa soqta pachak pusaq chunka isqon waranqa pisqa pachak tawa chunka kinsa-yoq*.

Before the next example, I would like to point out that, although it seems like the expression for 7,689,543 is long and cumbersome, this does not necessarily reflect on the efficiency of the language (in the case, Quechua). Indeed, if we write out the value 7,689,543 in English, we have *seven million six hundred eighty-nine thousand five hundred forty-three*, which is also a mouthful. Also, notice that the Incas use a *decimal* system.

Example 10.6 In this example we start with the Quechua expression and determine the value. The expression will be: *isqon chunka waranqa waranqa tawa pachak pusaq chunka iskay waranqa hoq pachak qanchis chunka kinsa-yoq*.

To construct the value, let us start on the left, with *isqon chunka waranqa waranqa*. This is a value of $9 \times 10 \times 1{,}000{,}000$, which we would write as 90,000,000. Next, we look for the term of thousands, which is *tawa pachak pusaq chunka iskay waranqa*. This expression amounts to, in the structure outlined above,

$$[[4]\times\{100\}+[8]\times\{10\}+(2)]\times\{1000\}=482{,}000.$$

So far, we have 90,482,000. To finish, we look at *hoq pachak qanchis chunka kinsa-yoq*. This expresses the value of 173. Thus, the value of *isqon chunka waranqa waranqa tawa pachak pusaq chunka iskay waranqa hoq pachak qanchis chunka kinsa-yoq* is 90,482,173.

I hope you now have an understanding of the mechanics of the Inca number system via expressions in Quechua. The next step is to get an idea of the cultural side of Inca counting.

CULTURAL ASPECTS OF THE INCA NUMBER SYSTEM

In Inca culture, as we have seen with other cultures in this book, counting is closely tied to important cultural aspects. Let us look at this aspect of Inca mathematics more closely. First, the *order* of objects can be important. Such is the case for counting ears of corn. Ears of corn on a stalk are carefully named in the order in which they appear, with the first representing the "mother" and others being defined in accordance to the first as offspring of the mother. See Figure 10.2.

Also, the concept of *pairs* is as important as the counting itself. It turns out that the Incas view the property of being even in a count as extremely

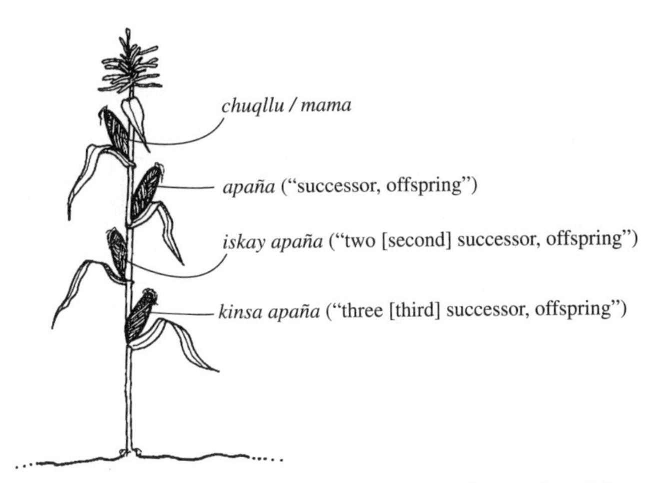

Figure 10.2 "Birth order" and naming of ears of corn. (Reproduced from *The Social Life of Numbers: A Quechua Ontology of Numbers and Philosophy of Arithmetic*, Gary Urton with the collaboration of Primitivo Nina Llanos, copyright © 1997, figure 3.3, p. 86. By permission of the University of Texas Press.)

important. In fact, we could go so far as to say that odd numbers are considered as incomplete pairs. There are many instances in which one counts pairs in Inca society. The cultural impetus for this viewpoint is that the Incas constantly strive to have proper order in their world. Such order extends to contexts like social situations. In this context if a person has done something inappropriate, then something must be done by the person or by the community so that proper social order is regained. This can be thought of as completing an incomplete pair. Finally, there are contexts in which quantities are not allowed to be counted. One such context is money: Money is not to be counted in traditional Inca culture. In the exercises below you will be asked to contemplate these cultural contexts of Inca counting.

Our next topic is Inca artwork. We will look at the mathematical properties, such as symmetry, as well as the cultural contexts and implications of the artwork.

INCA ART AND DECORATION

Let us consider the situation of a group of Inca women who are weaving, similar to situations we have seen in Chapters 5 and 9. Figures 10.3a and b show some examples of weaving patterns of the Inca. In Figure 10.3a, for example, we can observe many types of symmetry. One such symmetry is *horizontal*. Similarly, the pattern on the left half is exactly repeated on the right half, so the figure has, except for small variations, *vertical* symmetry.

Figure 10.3 (a)–(b) Bolivian textile pattern. (Courtesy of *Bolivian Indian Textiles, Traditional Designs and Costumes*, Tamara E. Wasserman and Jonathan S. Hill, Dover Publications, 1981, plates 11 and 17.)

Now let us look at Figure 10.3b. Except for slight displacements of some of the smaller parts of the figure, it has horizontal symmetry. It does not have vertical symmetry, however.

As we know, symmetry can appear in contexts other than weaving, such as in ceramics or strip patterns. You can ponder the symmetry properties in Figures 10.4 and 10.5. In Figure 10.4, the Ica culture represents a culture that was under Inca control. The Moche existed before the rise of the Incas, though the Chimú culture that did exist during the Inca era was at least partially rooted in Moche culture. At any rate, you can see that in Figure 10.5, on the top of the facial part of the image, there is a kind of repeating pattern. It could be considered as a strip pattern, though we do not know if it continued on the back side of the figure. Moreover, there does not seem to be the tendency to use patterns with many symmetry properties (i.e., that have horizontal, vertical, and 180° rotational symmetry). Rather, the patterns seem to be asymmetric. This could imply a difference in mathematical tendencies between the

Figure 10.4 Ica design patterns. (Reproduced from *Pottery Style and Society in Ancient Peru: Art as a Mirror of History in the Ica Valley, 1350–1570*, Dorothy Menzel, by permission of the University of California Press, 1976, part of plate 26.)

Figure 10.5 Moche ceramic vessel. (Courtesy of *Pre-Columbian Design*, Alan Weller, Dover Publications, 2008, plate 077.)

Moche culture and the later Inca culture. Although we would need to study many more pieces of artwork of both cultures, we can speculate that this is an example of how we can use cultural mathematics to understand cultural differences.

Example 10.7 In Figure 10.6, indicate the symmetry properties.

You can see that it appears as though there is vertical symmetry; however, some of the angled stripes cross over others, so in fact, there is neither vertical nor horizontal symmetry in this example. On the other hand, if you rotate the image 180° and carefully note how the angled stripes overlap others, you will see that the figure does have 180° rotational symmetry.

Example 10.8 In Figure 10.7, determine to which of the seven symmetry groups the figure belongs.

The interesting piece of art shown in Figure 10.7 is an example of mathematics in artwork made from clay, as opposed to cloth. Think back to the three-dimensional aspect of art from baskets and ceramics that we discussed near the end of Chapter 5. The pattern is split into two parts, but we can

Figure 10.6 Inca textile. (Reproduced from *Textiles of Ancient Peru and Their Techniques*, Raoul d'Harcourt, by permission of the University of Washington Press, 2002, plate 55C.)

Figure 10.7 Peruvian pottery. (Courtesy of *Appleton's American Indian Designs*, Le Roy Appleton, Dover Publications, 2003, image 709.)

consider it to be repeating in the sense of a strip pattern. If we consider it to be a strip pattern from the top to the bottom, then the pattern is of type 1 (see Chapter 5, page 82), because the plant-like symbols satisfy only translation, as taken in pairs of "plants." If we look at the strip pattern along the top of the figure, we can see that it satisfies vertical, horizontal, glide reflect, and so on, with alternating colors. For example, the small squares in the middle of the pattern alternate between black and white (at least in this black and white figure). Hence, we can say that this part of the pattern is type 7 with alternating colors. In particular, note that even numbers are represented, in the form of pairs. Several parts of the pattern appear directly in pairs or in groups of four. The small squares appear in groups of seven; however, the patterns of seven appear four times for an even number of 28 small squares (56 total if the same pattern is repeated on the back of the figure).

The above is good practice for reviewing symmetry patterns, but now let us go back to the cultural part of art. We can ask the question: What can we deduce from the symmetry properties used in this particular cultural group? In the case of the Incas, we see the repeated preference for extensive symmetry. You will witness this in the exercises, where you will look at further examples of Inca art and the symmetry properties in those pieces of art. This preference indicates a strong sense of order and precision in Inca culture and mathematics. It corresponds with our general knowledge of the Incas as being a culture that kept careful records (this will be detailed in the section on Inca number symbols coming up later). We previously observed the importance of pairs in Inca culture, and we can see how this is manifested in Inca artwork. In Figure 10.3b, for instance, the leftmost part of the pattern has alternating colors. We can view this as a complementary pair. Another example is in Figure 10.8. Notice that the figure has vertical symmetry, with alternating

Figure 10.8 Peruvian design. (Courtesy of *Appleton's American Indian Designs*, Le Roy Appleton, Dover Publications, 2003, image 666.)

colors. This expresses the idea of pairs in the form of two colors alternating in the artwork.

In our context of cultural mathematics, we can observe some mathematical properties in Inca weaving. In Figure 10.9 we see that thread counts are done in pairs or blocks of 10 and 20 and that in general what is being counted are values of *pairs* of numbers of threads. Notice also the influence of the decimal system in Inca counting. The complex geometric or animal motifs are created by a process called *pallay*, which is a Quechua word that means "to pick up."

I will make some final comments on the cultural aspects of Inca artwork. Master weavers of traditional Inca culture are called *Mamas*. The word *Mama* is a Quechua word that refers to an expert weaver (and does not imply

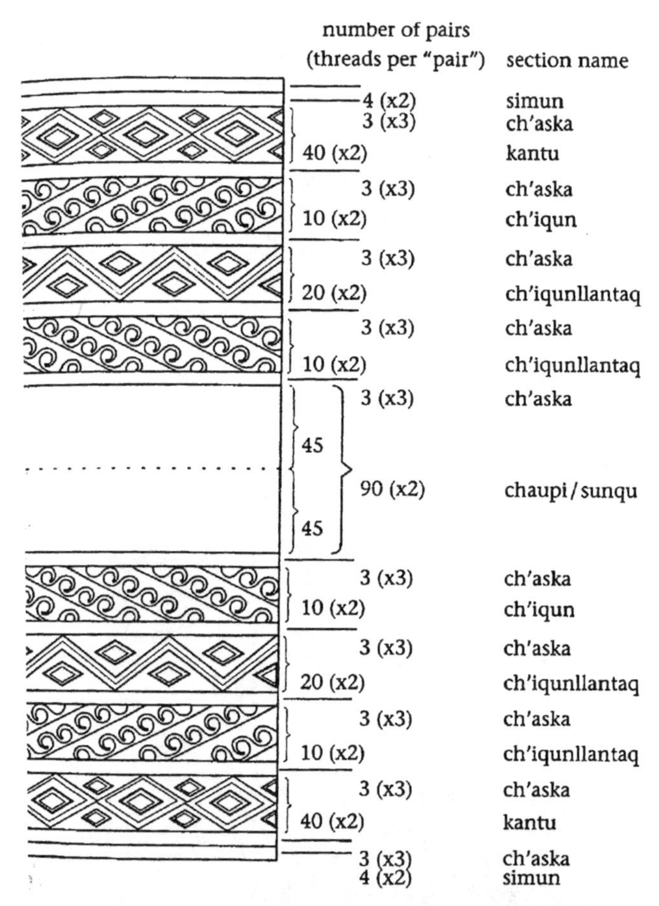

Figure 10.9 Warp counts in Inca weaving. (Reproduced from *The Social Life of Numbers: A Quechua Ontology of Numbers and Philosophy of Arithmetic*, Gary Urton with the collaboration of Primitivo Nina Llanos, copyright © 1997, figure 4.6, p. 121. By permission of the University of Texas Press.)

"mother"). *Mamas* are women who most likely started weaving when they were young girls and have reached a high level of expertise in weaving. As such, they are treated with special respect. Their abilities in counting, in understanding patterns of symmetry, and in geometry are part of that expertise. The cultural mathematics aspect of this situation is this: If we asked one of the women to explain geometry or symmetry properties in terms of lines, rotations, polygons, and so forth, her explanation probably would not be in terms that make sense to us. Yet the *Mamas* clearly understand these mathematical concepts; the difference is that their understanding comes from the perspective of weavers who want to create patterns for a cultural reason, not for the sake of practicing mathematics. Think back about how we have seen this before with the art of other cultures. Finally, notice the role of gender: Here we see women with mathematical skills being highly respected for their work.

Our next topic is Inca calendars.

INCA CALENDARS

The Incas had two calendars: a solar calendar of 365 days and a ritual calendar based on twelve lunar cycles. Recalling that a lunar cycle is approximately 29.53 days, twelve lunar cycles comes to about 354.4 days. This value is about eleven days short of a solar year of 365 (or 365.2422 in Western terms) days. It is not clear how the Incas accounted for this difference. The Spanish chroniclers indicated that the Incas adjusted their calendars near the spring solstice of the Southern Hemisphere (usually December 21 in the Western calendar), but the details of how the calendar was adjusted are not yet known. There are two proposed explanations. One is that the Incas inserted a short cycle of eleven days to make up the difference. The other is that they rounded the lunar cycles to thirty days.

What is known is that the Incas were good astronomers and had accurate information about the cycles of the moon, sun, and Venus, and knew when solstices and equinoxes occur. To date, there is no archaeological example of a physical calendar (e.g., made of stone or wood) that they might have used, and perhaps they did not have any such physical calendars. However, some stone structures in the city of Cuzco are related to Inca calendar considerations, as we will see now.

Throughout the city of Cuzco there are remains of about 400 markers, called *huacas*, along imaginary lines called *ceques* (also *zeq'e*). These lines originate from the center of Cuzco. Figures 10.10 and 10.11 show some details of the system. The *huacas* had ritual and social significance. Some had connections with astronomy. In particular, the Incas used several of the *ceque* lines as part of astronomical observations such as that of the June solstice. The *ceques* and *huacas* were crucial components of Inca religious ceremonies; in fact, the *huacas* are more accurately described as sacred shrines.

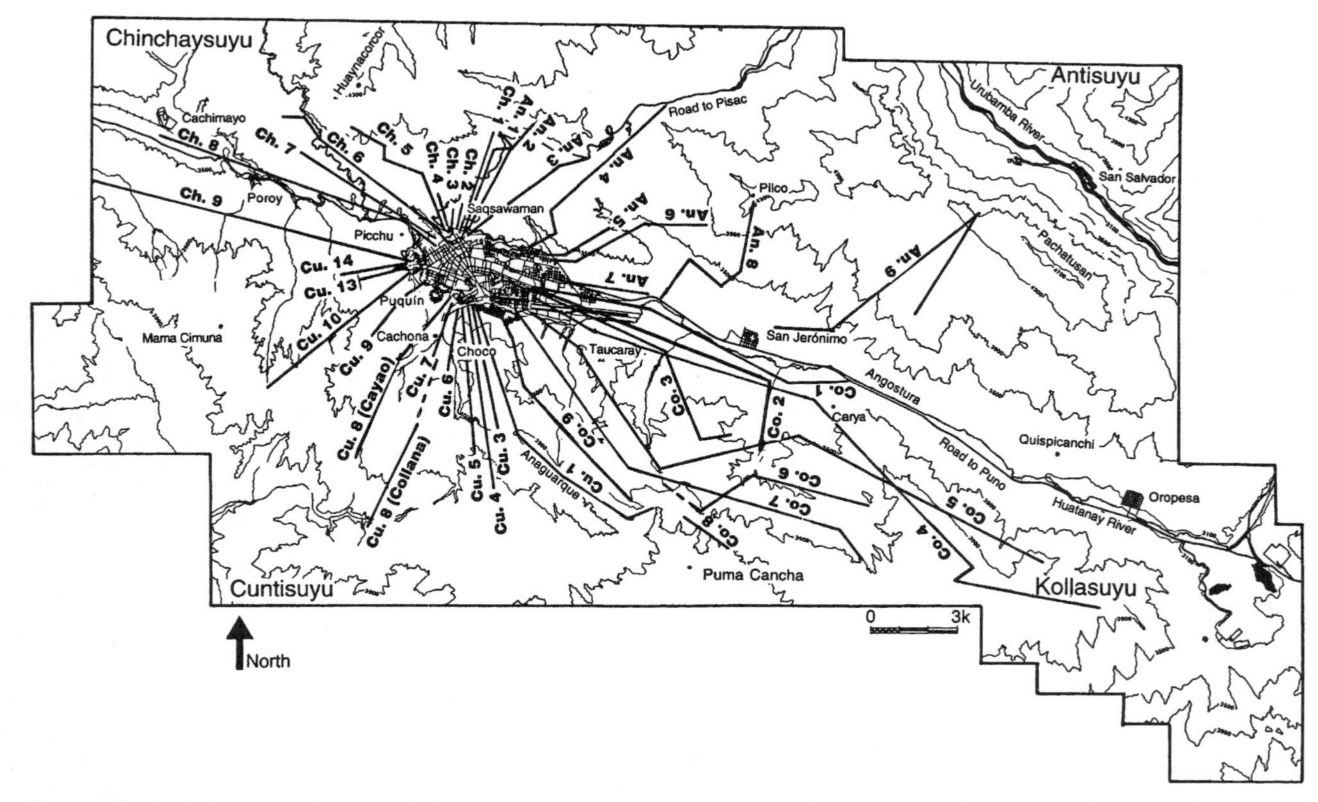

Figure 10.10 Schematic diagram of Cuzco's ceque system. (Reproduced with permission from *The Incas* (*Peoples of America*), Terence D'Altroy, John Wiley & Sons, Ltd., 2003, figure 7.1, p. 160.)

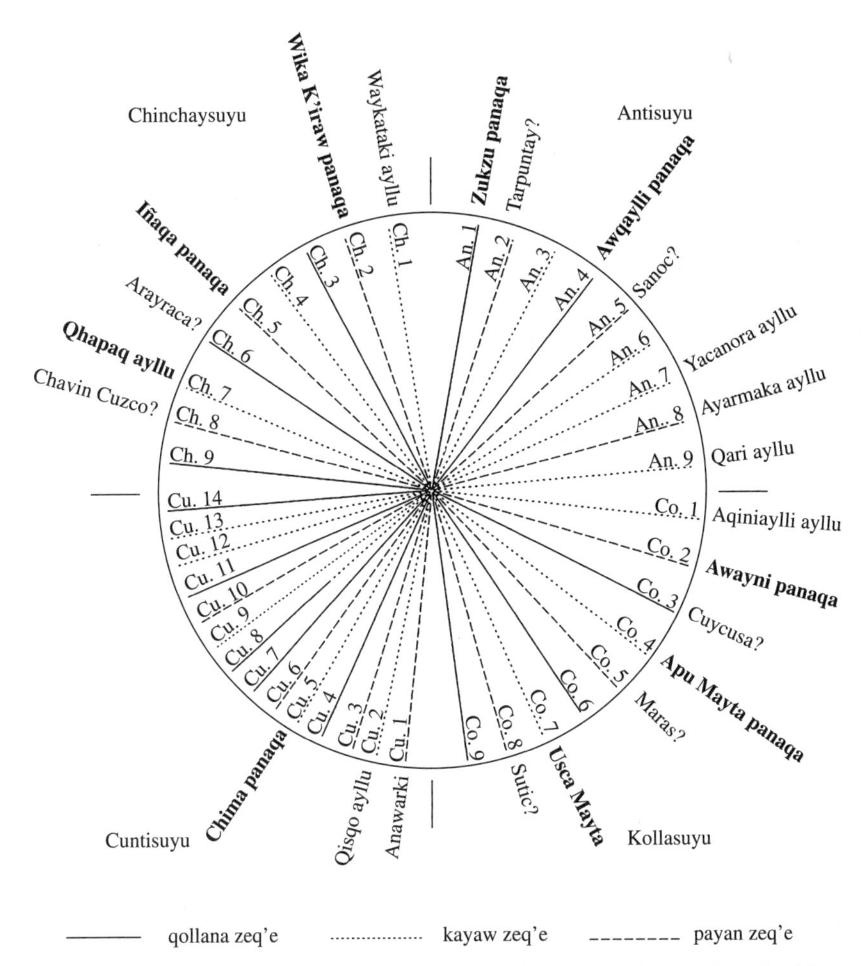

Figure 10.11 Schematic diagram of Cuzco's ceque system. (Reproduced with permission from *The Incas* (*Peoples of America*), Terence D'Altroy, John Wiley & Sons, Ltd., 2003, figure 7.2, p. 161.)

For more information about the Inca calendar, see Urton (1981) and D'Altroy (2003: chapter 7). Meanwhile, it is time to consider symbolic representations of Inca mathematics, the topic of the next section.

INCA NUMBER SYMBOLS: THE *QUIPU*

We now wish to look at the number symbols of the Inca. We will see that it is very different from other number symbols we have seen, such as those of the Egyptians, the Mayans, and the Otomies. In fact, the Incas did not have specific symbols for specific values, like the Mayans would use a bar for the value five, for example. What the Incas had was a sophisticated system of communication that was expressed as numerical data on knotted strings, called a *quipu* (pronounced "KEE-puu," and also written *khipu*). *Quipus* had existed for a long time before the rise of the Incas, with some as old as 5,000 years. On the other hand, the Incas made use of *quipus* in a big way by using them as their principal tools for storing numerical and other information. By arranging the knots and strings in certain ways, the Incas could represent many different kinds of numerical data. Our goal in this section is to understand some basic ideas of *quipus*.

As we now know, the Incas counted with a decimal system. We have also seen that they had a large territory to control. Moreover, it is well known that the Incas exacted tributes (taxes) from the groups they controlled and kept careful records of their tributes and governmental operations. The Incas kept track of such information by carefully tying knots on strings to represent the numerical values. Blank *quipus* were first formed, then filled with information in the form of knots on the cords. Most knots were tightly tied, which implies that *quipus* were intended to be record-keeping devices, not counting devices. A *quipu* consists of several types of cords. There is always a *main* cord, to which other, *pendant* cords are tied. There can be cords tied to the pendants called *subsidiary* cords, and there can be *top* cords and *dangle-end* cords. *Quipus* can have just a few cords, or up to thousands of cords. Some of the cords, such as top cords, often have totals of values of other cords as their values.

We can safely say that the Spanish conquerors and even most of the chroniclers such as Cieza de León, who described Inca culture, did not understand the mathematical structure of *quipus*. It was not until 1923 that Leland Locke uncovered the basic structure of the *quipu*. He found that there are three basic types of knots on a *quipu*, each representing a different kind of value, such as powers of ten, digits, and the number one. Most of our discussion below will highlight mathematical properties of *quipus*; however, it is important to note that the mathematics is only a piece of the picture. *Quipus* were used not just for keeping records, but also as part of the retelling of historical events and even song. See Quilter and Urton (2002) for more information about these cultural aspects. Moreover, there are more considerations regarding *quipus*, such as the orientation of the knots, the distribution of colors, and the types of materials used in *quipus*.

Take a look at Figures 10.12–10.14. An exhaustive study of *quipus* was undertaken by Marcia Ascher and Robert Ascher, starting in the 1970s (see, e.g., Ascher and Ascher, 1981; Ascher, 1983, 1986). Guamán Poma de Ayala (1980) made many drawings of *quipus*, and we can see four such drawings in Figure 10.15.

Now let us go deeper into the details of the three types of knots: *Simple* knots represent powers of ten. *Loop knots* (also called long knots) have

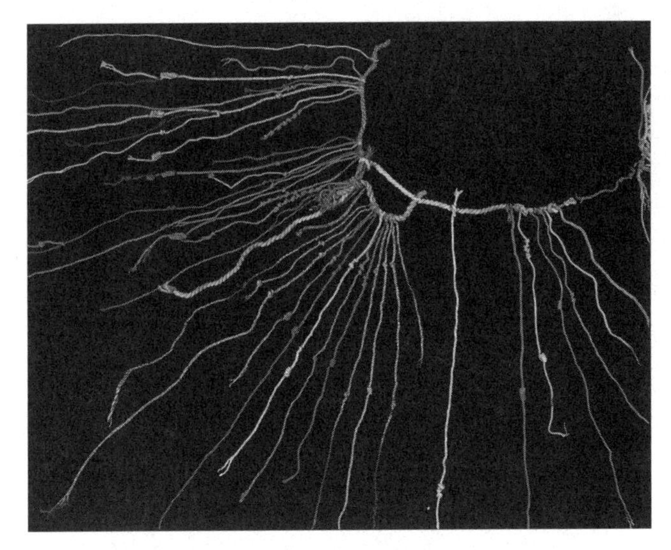

Figure 10.12 *Quipu* image. (Reproduced from catalog no. A289613-0, courtesy of Department of Anthropology, Smithsonian Institution.)

Figure 10.13 *Quipu* image. (Reproduced from catalog no. A365240-0, courtesy of Department of Anthropology, Smithsonian Institution.)

several loops that represent digits between two and nine, and *figure eight* knots represent the number one. The spacing of knots relative to other knots indicates value with respect to other knots, so in this way the Inca system is *positional*. That is, if there is a group of knots and then a noticeable space followed by another group of knots, then these groups of knots represent values multiplying distinct powers of ten, or possibly two distinct values. In addition, the *color* of the strings is important. Different colors can represent different types of data. Organization using color is comparable to different mathematical notations we use for variables. In Western mathematics, for example, instead of denoting real number quantities by x and vector-valued variables by $\mathbf{x}$, we could just as well distinguish those quantities by the *color* of the variable. So,

Figure 10.14 *Quipu* image. (Courtesy of Marcia Ascher.)

we could have a system in which, say, a green x refers to a real number while a purple x refers to a vector.

Apart from using a positional numbering system, the Inca used *quipus* to express the concept of *zero.* In fact, there are three ways the Incas could express zero using a *quipu*. One way is that a *quipu* could have a blank cord representing the idea that there is no quantity of the objects represented by the cord. Another way is that a *quipu* could have a missing cord for a quantity. As an example, consider a *quipu* that represents the amount of tribute (taxes) paid per household, for several communities. In one community it could be that there is a household that did not pay any tribute. In this case the *quipu* could have a blank cord for that house. In another community it could be that there is an area consisting only of agricultural land and storage buildings, but having no houses on it. In this case, the *quipu* could have a blank space with no cord on it to represent the tributes of that area. The third way to express zero is to leave a blank space between powers of ten; that is, to have a *place holder* of zero. This is done in Western decimal notation by putting the zero symbol in the appropriate place. For example, for the value of five hundred seven, we put a zero between the five and seven (507), to distinguish it from 57. At this point, we can learn more about *quipus* by looking at some examples.

For the notation for the different knots, I will use a lowercase "x" for simple knots, the symbol "8" for the figure eight (ones) knot, and a repeated loop for

Figure 10.15 Four drawings of *quipus.* (a) *Quipu* and dot figures; and (b)–(d) *quipus.* (Reproduced with permission from *El Primer Nueva Corónica y Buen Gobierno*, Felipe Guamán Poma de Ayala, Siglo XXI Editores, 1980, pp. 332, 330, 746, and 829.)

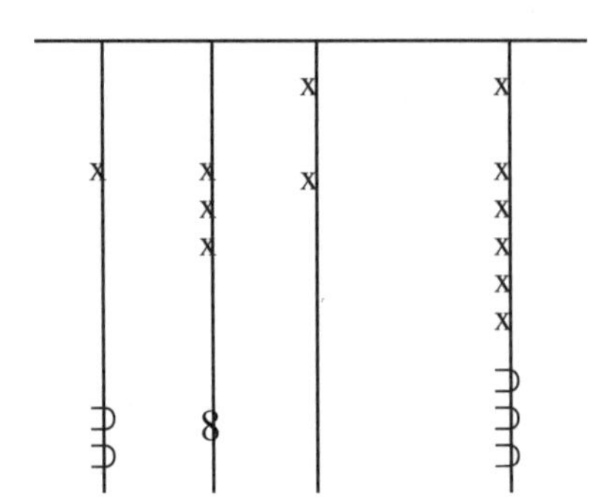

Figure 10.16 *Quipu* (see Example 10.9).

the values two through nine that looks like this: ⊃. Thus, for example, the value of seven would have the symbol ⊃ repeated seven times.

Example 10.9 An Inca farming family has baskets for keeping their crops. Suppose that they harvested 12 baskets of quinoa (a grain), 31 baskets of corn, and 110 baskets of potatoes (of various types). Construct a *quipu* that represents this data, including a cord that has a total value.

For the quinoa cord we need one simple knot in the tens place, then a two loop knot in the ones place. For the cord representing corn we need three simple knots in the tens place, then a figure eight knot in the ones place. For the cord representing potatoes, we need one simple knot in the hundreds place, one simple knot in the tens place, then the rest of the cord would have no knots to represent zero as a placeholder. Finally, for the total, we have 153 (add the values of the first three cords), so the fourth cord would have one simple knot in the hundreds place, five simple knots in the tens place, and a three loop knot in the ones place. A diagram of the *quipu* to record these quantities is shown in Figure 10.16. A few observations can be made. First, notice how the value of zero is expressed as a blank portion of a cord, indicating a placeholder use. Second, note how the positions of the knots play a role in the values. In particular, the simple knots have to be aligned pretty closely, because otherwise there would be confusion about which power of ten a simple knot represents.

The following is another example of information on a *quipu*.

Example 10.10 Consider the diagram of a cord of a *quipu* shown in Figure 10.17. The horizontal line segment represents the main cord. What can you conclude about this pendant cord?

This cord diagram seems to contradict the format of putting first simple (x) knots for powers of ten followed by either figure eight (8) or loop (⊃) knots

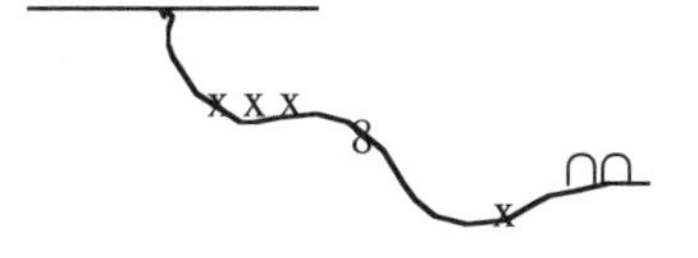

Figure 10.17 *Quipu* cord.

because there is a simple knot that appears after the figure eight knot. In fact, what we see on this cord is that there are *two* values on the cord. The first is 31, from the three simple knots followed by the figure eight knot, then a value of 12, shown as one simple knot followed by a two-loop knot. The gist of this example is that it is possible to exhibit even more complicated information on *quipus* by putting several values on each cord. Let us try another example of a *quipu* with two values on each cord.

Example 10.11 Construct a diagram of a *quipu* with three pendant cords such that one cord has values 42 and 15, the second cord has values 100 and 31, and the third cord has values 450 and 0.

To construct this *quipu* we will have to be quite careful about spacing. Here we go: On the first cord, we put in four simple knots followed by a two-loop knot, for the first value. For the second value we need a simple knot at the tens place and a five-loop knot. For the second cord we put in one simple knot in the hundreds space followed by blank space for zero, then three simple knots in the tens space followed by a figure eight knot. The three simple knots must be at the same level as the simple knot for the value of 15 on the first pendant cord. For the third cord, we put in four simple knots at the hundreds space followed by five simple knots in the tens area, followed by space where the loop or figure eight knots would go, and then more space for the second value. If our knots for some values are not spaced well enough to clearly show the powers of ten (e.g., we do not want the knots for 42 to look like a value of 420), we will have to "retie" (i.e., redraw) the knots to adjust for spacing. The Incas probably looked over the magnitudes of values (or potential magnitudes) before setting up a *quipu*, because to retie many knots would be too laborious to be practical. The final diagram is shown in Figure 10.18. While you are looking at the multiple values on the cords, take notice of how the value of zero has been expressed in this *quipu* as well.

The examples above show us that there are several parts of a *quipu* that give us information, including the placement of cords, positions and types of knots, and string colors. It turns out that there are about eight natural colors for use in *quipus* and that color is a very important part of the structure of a *quipu*. There is more than meets the eye with cords, too. Once the material for a *quipu* is chosen (e.g., cotton), the fibers are spun into a single-ply string

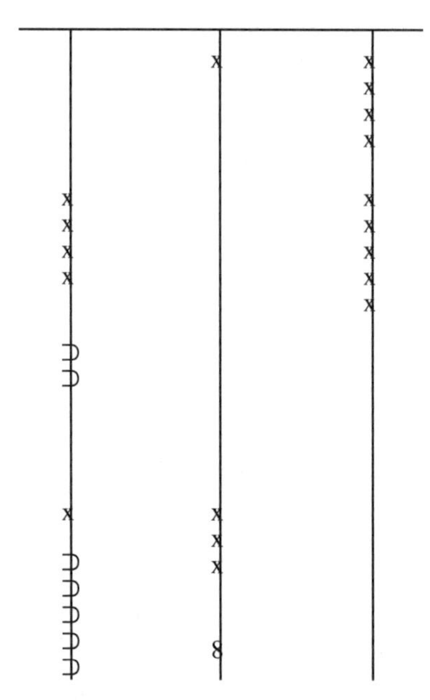

Figure 10.18 *Quipu* diagram (see Example 10.11).

having either an “S” pattern or a “Z” pattern. Then those single plies are doubled together, and sometimes doubled more than once. The final cord can be created by spiraling in a single Z pattern color. Mixing color leads to the creation of *quipu* cords that have multiple colors intertwined in their appearance. By embedding information in the color and plying scheme, we realize that a *quipu* can hold a tremendous amount of information. Moreover, it may seem in the final *quipu* that there is some haphazard spinning together of threads to make these cords; however' the process is *ordered and precise*. Recall the ordered, precise geometric symmetry patterns from the section on Inca art. Here is an explicit example of cultural mathematics: We have made a cultural observation using mathematics, not by looking at numerical computations, but rather by comparing patterns in weaving with types of knots and plying in threads.

The cord, color, and knot structures of the *quipu* allows for many levels of complexity. *Quipus* can be made even more complex in other ways. One way is to group colors, while another way is to associate groups of cords by spacing. Combinations of spatial grouping and color coding are also possible. For example, a *quipu* could be organized with twelve pendants that have four colors repeated three times, each color representing some distinct aspect of the information in the *quipu*.

We can see that there is a lot of flexibility with regard to organizing information on a *quipu*. In fact, the *quipu* can be considered as a representation of mathematical *symbols* whose type, location, and context also reveal non-numerical information. Thus, a *quipu* can be very sophisticated, containing a large amount of information!

Quipus were made by humans, of course, so let us look into that aspect of *quipus*. In pre-Conquest Inca, there were "*quipu* makers." The Quechua word for *quipu* maker is *quipucamayuq*. The *quipucamayuqs* were experts at constructing and interpreting *quipus*. They had to have significant mathematical knowledge and had to be able to explain the information on complicated *quipus* to Inca royalty. Moreover, they often included other elements of interpretation such as songs or historical stories. This is an instance of mathematicians having high social status.

Another way to arrange a *quipu* utilizes a Western mathematical concept known as a *tree structure*. In this case, levels of information with increasing complexity can be obtained by attaching more levels of subsidiaries of pendants to other subsidiaries, continuing this process to as many levels as desired.

Example 10.12 Suppose we are considering the hierarchy of a government system that is organized by a fixed management structure in each district. That is, each district has the same structure, given by the tree structure depicted in Figure 10.19. Each district has an executive position E and two management positions, M_1 and M_2, with M_1 having four subordinates and M_2 having two subordinates. Next, suppose we would like to log the number of hours worked in eight districts over a span of five years. How would we express this information on a *quipu*?

First, we note that there will be many copies of the tree structure shown, one for each district, and within each district, there will be a tree for each of the five years. The total will be eight trees repeated five times for a total of 40 trees. As for the cord structure, we could use specific colors of cords to denote each year, and the placement of each tree could be ordered from left to right starting with the first district.

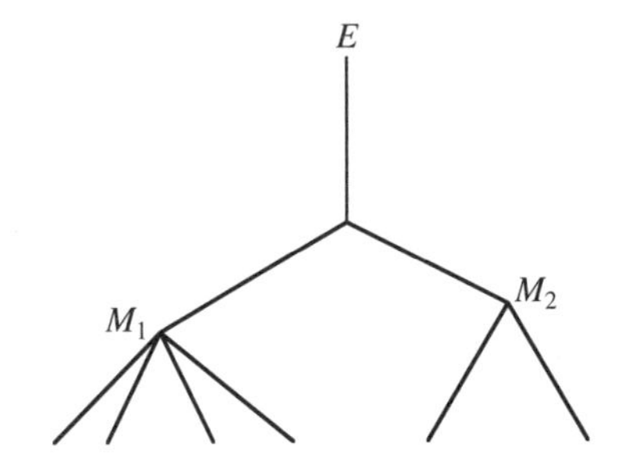

Figure 10.19 Tree structure in a *quipu*.

In the next section we will learn how to interpret data from real *quipus* that have been investigated.

INTERPRETING THE ASCHERS' *QUIPU* DATA

As mentioned earlier, Marcia Ascher and Robert Ascher pioneered the cataloging of *quipu* data for about 200 *quipus*. In this section I will describe some basic ideas of their data collection so that you will be able to interpret their work on *quipus*. I have left out some of the finer details (such as official codes for colors), but you can find those details in Ascher and Ascher (1981).

Figure 10.20 shows part of the data of one of the *quipus* studied by the Aschers. The figure represents only the first page of their collected data for that *quipu*; however, there is sufficient information for us to see how the data are organized.

Example 10.13 Determine the values of the cords listed in the partial description of *quipu* shown in Figure 10.20.

In going through this example, we will see how the information is presented by the Aschers. At the top of the page there is a general description of where the *quipu* was located when it was examined and a summary of how the pendant cords are arranged. In this example you can see that there are five groups of pendants, with the number of pendants in each group being four, six, eight, six, and six, respectively.

After the general summary there are columns of information with several categories: *Cord*: The cords of the *quipu* were numbered by the Aschers. Next is *Knots* (*no., type, position*): This refers to the types and placements of knots on the cord listed in the first column. The codes for the knot types are "*s*" for *simple*, "*L*" for *loop*, and "*E*" for *figure eight*. The term "*no.*" refers to the number of knots of each type. The position number refers to the number of centimeters from the main cord. Then there is *Length*: The total length of the cord. Next appears *Color*: There are official names of types of colors whose abbreviations were used by the Aschers. See Ascher and Ascher (1981) for more information about color labels. The next category is *Value*: The Aschers used the information on the types of knots and their placements to determine the value on the cord. Finally, there is a category of *Subsidiaries*: If there were no subsidiaries tied to the pendent cord, this part is left blank. Otherwise, information on knot types and spacing is given here for the subsidiaries.

Let us look at cord numbered "1" to review this information. On that cord there are four loops, starting from 13.0 centimeters (denoted *cm*) from the main cord. The total length of the cord is 36.0 centimeters, and the cord was of the color white (indicated by the letter *W*). The value—which we already

QUIPU AS210

Museum identification: No. 111407 (Field Museum of Natural History, Chicago)

Main Cord: color B

$ 1.0 cm: group of 4 pendant cords (1–4), then space of 2.5 cm
4.5 cm: group of 6 pendant cords (5–10), then space of 2.0 cm
8.0 cm: group of 8 pendant cords (11–18), then space of 2.5 cm
12.5 cm: group of 6 pendant cords (19–24), then space of 1.0 cm
15.0 cm: group of 6 pendant cords (25–30), then space of 64.0 cm
81.0 cm: end ¢

Cord	Knots (no., type, position)	Length	Color	Value	Subsidiaries (no., position)
1	4L(13.0)	36.0¢	W	4	
2	4L(12.0)	35.0¢	KB	4	
3	2L(13.0)	28.0¢	B	2	
4	1E(13.0)	39.0¢	B	1	
5	2s(4.5) / 4L(12.5)	25.5¢	W	24	1 : 0.5
5s1	1a(4.0)	32.0¢	KB	10	
6	1a(5.5) / 1E(13.5)	29.5¢	B	11	1 : 0.5
6s1	4L(14.0)	35.0¢	B:RG	4	
7	5L(13.0)	33.0¢	B	5	
8	5s(5.0)	36.0¢	B	50	

Figure 10.20 *Quipu* AS210 (partial image). (Reproduced from *Code of the Quipu Databook I & II*, Marcia Ascher and Robert Ascher, University of Michigan Press and Cornell University Archives, 1978 & 1988, item AS210. Courtesy of Marcia Ascher.)

know how to calculate as four loops—is four. There was no subsidiary cord tied to Cord 1.

Now let us look at Cord 5. It has two simple knots, starting 4.5 centimeters from the main cord, followed by four loops that start 12.5 centimeters from the main cord. Cord 5 is 25.5 centimeters in total length, of color white, and it has a subsidiary. We can calculate the value by noting that it will be twenty (two simple knots) and four (four loops), so 24, as listed in the Value column. The notation "1; 0.5" in the Subsidiary column means there is one subsidiary and that the subsidiary is tied to pendant Cord 5 at a length of 0.5 centimeters from the main cord. You will notice that below the information for Cord 5 there is a row starting with the term "5s1," which refers to subsidiary one of

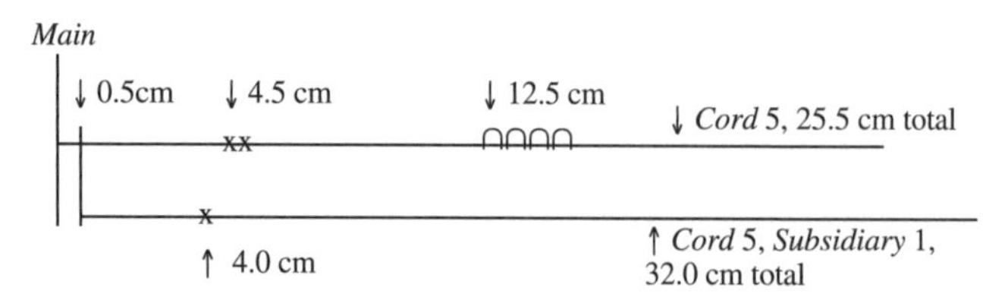

Figure 10.21 Diagram of Cord 5 and its subsidiary (approximately to scale) of *quipu* AS210.

Cord 5. If there were more subsidiary cords, each would have its own description here. Figure 10.21 shows a diagram of Cord 5 with its subsidiary, drawn horizontally.

You can go through a similar process to determine the values of the other cords, and you can look at the Values column to check your work. In the exercises you will have a chance to describe *quipu* information in a way similar to the above example, although for those examples the values of the cords are not given explicitly, and you will have to figure out the values.

Now I will present another observation using cultural mathematics. Recall from the section on Inca artwork that the precision and highly symmetric patterns seem to imply that the Incas had a cultural tendency for order and organization. This tendency appears in the *quipu* data as well, in the following way. Notice that on Cords 1, 2, 3, 5, 6s1, and 7, there are loop knots. The distance between these loop knots and the main cord varies between 12.5 and 14 centimeters; that is, there is less than 2 centimeters of difference between how far the loops are from the main (or, in the case of 6s1, from the pendant) cord. Similarly, the simple knots, on Cords 5, 5s1, 6, and 8, vary in distance from the main (or pendant) cord with values ranging from 4.0 to 5.5 centimeters. Again, this is less than 2 centimeters of difference. Although some *quipucamayuqs* were less careful than others (think of it as variation in neatness of handwriting; see Quilter and Urton, 2002, for more details), there is an overall tendency toward *order and organization* in the spacing and arranging of knots on *quipus*.

Example 10.14 In this last *quipu* example, I would like to discuss a particularly mathematical *quipu* that was investigated by the Aschers. It consists of two *quipus* tied together. Although simple in construction, the *quipus* contain surprising mathematical properties. We will focus mainly on the mathematical information, keeping in mind that there is more information that can be gleaned from the color and length of the cords, as well as the context (which we do not know) of when and why the *quipu* was made. See Figure 10.22.

First, note that one *quipu* has seven pendants and three subsidiaries (AS56), while the other has three pendants only (AS55), for a total of thirteen pendant

QUIPU AS55

Museum identification: No. 4A (Museo de Ica, Ica, Peru)

Main cord: color YB

*$ 2.0 cm: group of 3 pendant cords (1-3), then space of 33.5 cm.

36.0 cm: end ¢

Cord	Knots (no., type, position)	Length	Color	Value	Subsidiaries (no., position)
1	1s(2.0); 4s(9.0); 2s(16.0); 1E(21.0)	30.0¢	KB	1421	
2	2s(2.0); 4s(9.0); 2s(16.0); 7L(20.5)	25.5¢	YB	2427	
3	7s(9.0); 3s(15.5); 4L(17.5)	23.5¢	KB	734	

(a)

QUIPU AS56

Museum identification: No. 4B (Museo de Ica, Ica, Peru)

Main cord: color B

$ 3.0 cm: group of 7 pendant cords (1-7), then space of 17.0 cm.

21.5 cm: end ¢

Cord	Knots (no., type, position)	Length	Color	Value	Subsidiaries (no., position)
1	4s(5.5); 7s(15.0); 6s(23.5); 9L(31.0)	37.5¢	B	4769	2:10.0-10.5
1s1	1s(4.5)	24.5b	CB	100	
1s2	3s(5.0); 9s(13.5); 8L(21.0)	38.5¢	CB	398	
2	6s(14.0); 2s(21.0)	30.0b	BB	620(?)	
3	2s(5.5); 3s(13.5); 5s(21.5); 3L(30.5)	38.5¢	B:BB	2353	1:7.5
3s1	1s(6.0); 3s(14.0)	20.0b	CB	13?	
4	9s(15.0); 3s(22.5) 4L(32.0)	35.0¢	RL	934	
5	1s(5.0); 1s(13.5); 1s(21.5); 8L(32.0)	40.0¢	B	1118	
6	2s(5.0); 1s(13.5); 2s(22.0); 1E(32.0)	79.0¢	EB	2121	
7	7s(14.0); 5s(22.0); 6L(30.0)	37.0¢	BB	756	

(b)

Figure 10.22 *Quipus*: (a) AS55 and (b) AS56. (Reproduced from *Code of the Quipu Databook I & II*, Marcia Ascher and Robert Ascher, University of Michigan Press and Cornell University Archives, 1978 & 1988, items AS55 and AS56. Courtesy of Marcia Ascher.)

and subsidiary cords. Let us look at AS55. In the table of cord descriptions, Cord 1 has knot information given as 1s (2.0), 4s (9.0), 2s (16.0), and 1E (21.0), with the information organized in a way that we now can interpret after the previous example. Let us determine the value of Cord 1: The knot "1s (2.0)" is a simple knot, the highest on the cord. It is followed by a space after which appears the next group of four simple knots (9.0 centimeters from the top of the cord). Knowing what we do about the positional system of *quipus*, we deduce that the space between the first knot and the next group of four knots represents a difference in a power of ten. There are four groups of knots on this cord ("1s," "4s," "2s," and "1E"), and that tells us that "1s" multiplies 1000, "4s" multiplies 100, and so forth. The total value of the cord is therefore:

$$1 \times 1000 + 4 \times 100 + 2 \times 10 + 1 \times 1 = 1{,}421.$$

If we look at the Value column with respect to Cord 1, it confirms that the value is 1,421. Similarly, Cord 2 has value 2,427, and so on. You may wish to look at AS56 and cover up the Value column, then calculate the value of each cords to see if you get the same value.

Our interest is in the mathematical connections between the cords of these two connected *quipus*. Let us put six of the seven pendants from AS56 and the three pendants from AS55 in a table of values. I will organize the information in a manner similar to the Aschers' notation. That is, the notation p_{ij} will be used for AS56 values and q_{ij} for AS55 values, where i and j vary from 1 to 3. I will use standard matrix notation (for representing tables of information) in which the ij entry is what appears in the ith row and the jth column. For example, p_{23} is in the second row, third column. The format appears in Table 10.2, with the specific values listed in Table 10.3.

TABLE 10.2 Organization of *Quipu* Data from AS55 and AS56

Quipu	Value Notation		
AS56	p_{11}	p_{12}	p_{13}
AS56	p_{21}	p_{22}	p_{23}
AS55	q_{31}	q_{32}	q_{33}

TABLE 10.3 Some Actual Values of AS55 and AS56

Quipu	Values		
AS56	$p_{11} = 1620(?)$	$p_{12} = 2353$	$p_{13} = 934$
AS56	$p_{21} = 1118$	$p_{22} = 2121$	$p_{23} = 756$
AS55	$q_{31} = 1421$	$q_{32} = 2427$	$q_{33} = 734$

The Aschers looked very carefully at this *quipu* and found that the cord we are calling p_{11} was broken almost exactly at the place where one simple knot in the thousands place would likely be. By conjecturing that there was a simple knot at that location, $p_{11} = 620$ has a value more correctly given by 1,620. You will see in a moment that this value of 1,620 corresponds more logically to the rest of the *quipu* information than a value of 620.

By looking at the values it seems that the numbers in column 3, for example, are somewhat close in value. Take $p_{13} = 934$ and $q_{31} = 734$. Their ratio is $734/934 \approx 0.78586$. This is fairly close to $11/14 \approx 0.78571$. Coincidence? It turns out that with an accuracy within 0.4 percent (0.2 percent for most of the calculations), there are many approximations and relationships like this between values on these connected *quipus*. Some of those relationships are: $q_{32} = (34/33)p_{12}$, $q_{31} = (7/8)p_{11}$, and $p_{22} = (7/8)(34/33)p_{12}$. In fact, the fractions 11/14, 7/8, and 34/33 appear in several places in this *quipu*. Is there some connection between these fractions? Surprisingly, 7/8 is a rather accurate approximation to

$$\sqrt{\frac{11/14}{34/33}}.$$

Thus, we can conclude that the Incas probably knew what these values were. What their importance was, and in what context the values were put on a *quipu*, we do not know.

Here is one final observation: In this *quipu* example, fractions were formed by using one particular cord as numerator and another as a denominator. There are other possibilities for expressing fractions, such as using a top cord as the denominator. At this point you can probably think of some other ways that fractions could be expressed on a *quipu*.

I hope that these examples give you some idea as to how information was stored on *quipus*. The next topic has to do with how the Incas might have made calculations before putting the values on their *quipus*. We will see that the question of calculations is not so easily answered.

YUPANA

Yupana is a Quechua word that means *to count*. The *quipu* descriptions indicate that the Inca performed rather complicated calculations. Remember, the *quipu* was a record-keeping device and not used for calculating. So, as far as we know, the Incas were not using the *quipus* to determine values, only to store values. In light of Example 10.14, and knowing that some *quipus* had many hundreds of cords on them, we are led to the question: How did the Incas do their computations? The answer to this question is still not understood.

One possibility arises in one of the drawings of *quipus* made by Guamán Poma de Ayala, shown in Figure 10.15a. If you page back to that figure, you will see that there is a kind of a grid drawn in the lower left corner. The rectangular grid of solid and unfilled dots looks like it could be a kind of counting board; however, no explanation of the grid was given by Guamán Poma de Ayala. Certainly, there would be reason to believe that the grid has mathematical meaning because of its appearance with a *quipu*. On the other hand, there are some problems with concluding that the grid diagram represents an Inca counting device. For one thing, no archaeological artifacts representing the pattern of the grid have ever been found. In addition, because Guamán Poma de Ayala's work occurred after the Spanish conquest of Peru, the grid does not necessarily represent a device of Inca origin.

Nevertheless, it is possible to describe ways in which the grid could have been used for calculations. Several such interpretations of the grid can be found in Mackey et al. (1990). From the archaeological side, although no artifacts have been found that match the pattern in Guamán Poma de Ayala's drawing, there are stone artifacts of Inca origin that have different patterns. Images of some of these stone artifacts can be found in Mackey et al. (1990). There are descriptions by Spanish chroniclers of observations of Incas counting on these rectangular forms, but here we encounter the problem of accuracy. This is because when the Spanish described people of Inca culture moving pebbles or grains among the square places, we do not know if the Incas were counting values or playing a game. At the time of this writing, a conclusive explanation of the grid and stone artifacts still has not been given.

We have seen many aspects of Inca mathematics. Now you will have a chance to experience this material through some exercises.

EXERCISES

10.1 The Nazca lines were made by forming white lines in the desert. How do you think this was done? Keep in mind that many of the lines are very long, miles in length!

10.2 What was the European invasion of Inca territory that occurred before the arrival of the Spanish?

10.3 Name at least two obstacles that would make understanding Inca mathematics difficult.

10.4 Maria Reiche wrote books and articles about the Nazca lines during the 1950s and 1960s.

(a) What advantages did she have in accurately determining mathematical aspects of the Nazca?

(b) What disadvantages did she have in accurately determining the mathematical/cultural aspects of the Nazca?

10.5 Explain why we might initially expect Cieza de León to be a poor source of information about Inca mathematics.

10.6 **(a)** Explain why we might expect Guamán Poma de Ayala to be a good source of information about Inca mathematics.

(b) Explain why we might expect Guamán Poma de Ayala to be a poor source of information about Inca mathematics.

10.7 Suppose a child from Inca culture arrives in the United States. Explain several aspects of how that child might have difficulty with counting problems in school (including language and also cultural factors).

10.8 **(a)** Explain two ways the Inca could have adjusted their ritual and solar calendars so that they would coincide.

(b) Do the adjustments in part (a) represent natural, physical, or cultural cycles? Explain.

10.9 Why are there four main parts to the Inca ritual calendar?

10.10 What connection is there, if any, between the Inca calendars and Inca artwork? Explain.

10.11 Referring to Figure 10.9 on thread counts in Inca weaving: Explain the connection between thread counts and the Inca emphasis on pairs and even numbers.

10.12 Refer to the Guamán Poma de Ayala drawing of a *quipu* and dots figure (Fig. 10.15a).

(a) Explain why the dot arrangement might be related to Inca mathematics. On the other hand, explain why we are not guaranteed that this is genuine Inca mathematics.

(b) There is something inconsistent about the dot arrangement and Inca mathematics. What is that inconsistency? Give a possible explanation for the dot arrangement.

10.13 Describe a way that fractions could be expressed on a *quipu*.

10.14 It is known that the *quipu* was comparable to a spreadsheet, but not like a database. Explain how we can conclude this.

10.15 What numerical information of the Incas is expressed in the title of this chapter? Explain.

Calculations

10.16 For the Quechua expression *piqa pachak iskay chunka pusaq waranqa qanchis chunka tawa*, do the following:

(a) Write out the parts of the number in the [multiplier] {nucleus} (adder) format.

(b) Determine the number that is described by the expression.

10.17 Determine the values of the following numbers in the same way as in Exercise 10.16:

(a) *kinsa pachak tawa chunka qanchis waranqa iskay.*

(b) *soqta chunka piqa waranqa waranqa isqon pachak iskay chunka pusaq waranqa qanchis pachak soqta chunka hok.*

10.18 Construct the Quechua representation of the following values.

(a) 182,774

(b) 900,000

(c) 2,050,021.

10.19 Recall that a *supracycle* is when two cycles coincide. Often such days have cultural importance.

(a) Calculate the supracycle of the Inca solar and ritual calendars, assuming a lunar cycle to be 29 days. Indicate how many ritual cycles ("ritual years") and solar cycles (years) the supracycle represents.

(b) Calculate the supracycle of the Inca solar and ritual calendars, assuming a lunar cycle to be 30 days. Indicate how many ritual cycles ("ritual years") and solar cycles (years) the supracycle represents.

(c) Is there a way to calculate the supracycle of the Inca solar and ritual calendars using a value of 29.5 days for a lunar cycle? If so, determine the supracycle. Indicate how many ritual cycles ("ritual years") and solar cycles (years) the supracycle represents.

10.20 The artwork shown in Figure 10.23 is from the Chimú culture. Do the following:

(a) Determine which symmetry properties (vertical, horizontal, 180° rotational) the object has.

(b) Explain how the concept of even and odd are expressed in this artwork.

(c) Recall that the Chimú were an independent culture but were eventually controlled by the Incas. Indicate whether there are features of this artwork that indicate a difference in mathematical tendencies between the Chimú and the Inca. If there are also (or only) mathematical similarities with the Inca, explain that as well.

Figure 10.23 Pre-Columbian Chimú poncho. (Courtesy of the Lombard Museum.)

Figure 10.24 Pre-Columbian Inca artwork. (Courtesy of *Pre-Columbian Design*, Alan Weller, Dover Publications, 2008, plate 117.)

10.21 For the artwork shown in Figure 10.24, do the following:

(a) Determine the angle A for which the figure has $A°$-rotational symmetry.

(b) Explain how the concept of even and odd are expressed in this artwork.

Figure 10.25 Ancient Peruvian cloth. (Reproduced from *Textiles of Ancient Peru and Their Techniques*, Raoul d'Harcourt, by permission of the University of Washington Press, 2002, plate 17A.)

10.22 For the artwork shown in Figure 10.25, do the following:

(a) Determine to which of the seven symmetry groups the strip pattern belongs. Assume that there is a border on the bottom that matches the top border.

(b) Explain how the concept of even and odd is expressed in this artwork.

Exercises 10.23–10.32 involve making diagrams of varying types of quipus. *Be sure to use a consistent notation for the three kinds of knots. Remember spacing, and express yourself with color!*

10.23 Construct diagrams of *quipus* having the following properties:

(a) A *quipu* that has three pendant cords:

Cord 1 has values 28 and 41,

Cord 2 has values 902 and 55,

Cord 3 has values 21 and 300.

(b) A *quipu* that has two pendant cords: a subsidiary cord to Cord 2, and a *top* cord with a total of all values:

Cord 1 has values 82 and 41,

Cord 2 has values 0 and 333,

Cord 3 is a subsidiary cord to Cord 2 with values 600 and 201.

10.24 Consider a tax accounting system in which we have two states. Towns A, B, and C are in State 1, while Towns D and E are in State 2. Suppose the collected taxes from the states are to be recorded on a *quipu*, and we have the following data (in number of "tax items"): Town A contributed 88 items, B contributed 52 items, C contributed 451 items, D contributed 1,120 items, and E contributed 49 items. Construct a diagram of a *quipu* for this data, including subtotals by town and by state, and a total for both states. Be sure to *color-code the strings in a coherent way*.

10.25 Construct a diagram of a *quipu* that represents an inventory of farm animals with the following information: There are two farmers. One has 34 chickens, 41 goats, and no sheep. The second farmer has 233 chickens, 12 goats, and 6 sheep. Include a color scheme and a total value of all animals.

10.26 Describe how to construct a diagram of a blank *quipu* that represents an order form for a catalog or Internet clothing company. The structure of the *quipu* must include the customer's mailing address, including street numbers and zip code, credit card number, phone number, and dollar amounts of the purchase(s). Fill in your blank *quipu* with (fictitious, of course) information about a specific Internet order for some items.

10.27 Construct a diagram of a blank *quipu* that represents a local government in which there are four districts: Three districts have two counties each, and in each county there are five cities; the fourth district has three counties: County 1 has seven cities, County 2 has one city, and County 3 has ten cities. Be sure to use colors that represent similar quantities, and remember that this is to be a blank *quipu* with no data.

10.28 Explain three ways the concept of zero could be expressed on a *quipu*. Construct diagrams of examples of *quipus* that illustrate each of these three ideas.

10.29 Consider the situation of a *quipu* that represents crop harvests for several farmers in some region of Inca territory. Suppose Farmer A usually harvests corn, but this year there was no crop because of weather-related conditions. Suppose Farmer B lives at a high altitude where corn never grows.

(a) Explain how the harvest quantities for each farmer would be represented on the *quipu*.

(b) Explain how this indicates the Incas' understanding of the concept of zero.

10.30 Construct a diagram of a *quipu* that represents harvests, with 2,418 baskets of *maca*, 900 baskets of *kañiwa*, and 1,001 baskets of *kiwicha* (amaranth).

10.31 Construct a diagram of a *quipu* that represents harvests, with 601 baskets of yellow beans, 892 baskets of coffee beans, 1,300 baskets of cocoa.

10.32 Construct a diagram of a *quipu* that represents differing potato harvests, with 2,481 purple potatoes, 3,000 red cone potatoes, and 444 yellow/purple combination potatoes.

For Exercises 10.33–10.35 present data of quipus *similar to those studied by Marcia Ascher and Robert Ascher. Determine the value of each cord. Then reconstruct the* quipu *from the data. Keep in mind the color scheme!*

	Cord	Knots	Color	Value
		Number, type (position, in cm)		
10.33	1	4s (10) 6s (15) 5L (20)	B	_____
	2	7s (10)1E (20)	W	_____
	3	9L (20)	W	_____
	4	5s (10) 3s (15) 1E (20)	Y	_____
10.34	1	6s (10)	R	_____
	2	4s (5) 1E (20)	B	_____
	3	2s (5) 5s (10) 4s (15) 8L (20)	W	_____
	3s1	1s (10) 4s (15) 7L (20)	Y	_____
	(3s1 is tied 3 cm from the main cord.)			
10.35	1	8s (10)	G	_____
	1s1	5s (10) 2s (15) 1E (20)	W	_____
	1s2	4s (5) 1s (10) 6L (20)	W	_____
	2	–	B	_____
	(1s1 is tied 4 cm from the main cord, 1s2 is tied 6 cm from the main cord.)			

The following exercise requires some concepts of writing proofs.

10.36* Refer to Example 10.14 Prove that $(34/33)^2 \times (7/8)^6\, p_{12} = p_{21}$.

Essay and Discussion

10.37 Describe in some detail the obstacles in understanding Inca mathematics. Include factors of accuracy of information, cultural tendencies, and specific people considered to be reliable sources and why. You may wish to include the role of archaeology in this context as well—for example, how archaeological and anthropological methods have changed over the years and how such changes have affected the understanding of Inca culture (including Inca mathematics).

10.38 For the persons below, describe their importance to Inca mathematics:

(a) Marcia Ascher and Robert Ascher.

(b) Pedro Cieza de León.

(c) Felipe Guamán Poma de Ayala.

(d) Maria Reiche.

10.39 Write about or discuss a comparison of the mathematics of the Incas with that of the Otomies and/or the Mayans.

10.40 Write about or discuss a description of the different kinds of cords on a *quipu*. Describe how each type of cord could be used in a *quipu*.

10.41 Compare the four main calendar systems we have discussed: The Nuu-chah-nulth, the Mayan, the Otomi, and the Inca. Explain aspects such as which has more physical cycles, which has more natural cycles, and which has more cultural cycles. Also, explain the connections between each calendar and the cultural tendencies of the corresponding culture.

10.42 Write about or discuss possible Inca counting devices. Include the following aspects in your writing:

(a) Discuss the possibility that the "dot" drawing by Guamán Poma de Ayala represents an Inca counting device. How knowledgeable was he about mathematics? How do you know? What other factors are involved with this possibility?

(b) Investigate the possibility that the stone figures mentioned in this chapter represent Inca counting devices. What are the possible mathematical aspects of those figures? What other purpose might they have served? Mackey et al. (1990) is one place to begin such an investigation, for this part and for part (a) above.

(c) Discuss the possibility that *no* Inca counting device exists. What are the reasons why the Incas would have needed a counting device? What has been the development of counting devices in other cultures, including Western culture?

10.43 Write about or discuss a comparison between the mathematics of the Incas and that of the Otomies, discussed in Chapter 9.

10.44 *Research paper*: Choose a culture and investigate its mathematics as a case study such as we have done here. Keep in mind some of the resources, along with their advantages and disadvantages, that were discussed in the Introduction to Part II.

Additional Comments

- Most of the information for this chapter comes from the works of Marcia Ascher and Robert Ascher, two pioneers in the organization and description of Inca mathematics. If you enjoyed this chapter and want to learn more about Inca mathematics, the best place to start is by reading any or all of the works on the Incas and their mathematics written by them. In the Bibliography are listed the works with which I am familiar. Apart from those publications, you can find articles written by Marcia Ascher in Closs (1986) and by both Marcia Ascher and Robert Ascher in Quilter and Urton (2002).

- A rather complete description of cultural groups related to the Incas, as well as information about trade and cultural interaction between several Andean groups, can be found in Moseley (1992).
- If you are interested learning more about the Nazca lines, you can find information in Aveni et al. (1990). In fact, Anthony Aveni has written many works about ancient astronomy in the Western Hemisphere. Also, you can study most of Maria Reiche's original work on the Nazca lines in Reiche (1993).
- In Ascher and Ascher (1981: 3), you can find a detailed description of original works and translations of Cieza de León's work, and in the Bibliography here I have listed a reference to the works of Guamán Poma de Ayala from 1614, reprinted in Guamán Poma de Ayala (1980). He is particularly known for the many drawings of Inca culture that he made, a few of which were presented here.
- For more details about Incan astronomy and the *ceque* lines, see, for example, Urton (1981), Moseley (1992: 78–79), or D'Altroy (2003: chapter 7).
- All plant foods (such as quinoa) listed in the examples are native to the Americas and were cultivated by cultures such as the Incas. I have tried to use Quechua words when possible for the plant names (e.g., *kañiwa*). As of this writing, it is known that cultivation of plants in the Americas goes back to between 3000 and 2000 BCE. You may find it interesting to know that many of the foods we eat today were originally collected and/or cultivated by indigenous groups of the Americas. Such foods include corn, tomatoes, potatoes (many varieties were cultivated by the Incas as well as many other cultures), beans, squash, avocados, chocolate, coffee, and more. See Hays and Hays (1973).
- Some linguistic comments: Until the 1960s, it was commonly thought that Quechua originated and spread with Inca expansion. In Mannheim (1985, 1991), Stark (1985a, b), Weber (1989), and Urton (1997), you can find detailed explanations and references regarding the Quechua language. In particular, you will find that Quechua in fact originated in northern central Peru and later split into essentially two branches.
- Some general *quipu* comments: As Jeffrey Quilter tells us in the preface of Quilter and Urton (2002), the *quipu* is much more than just a simple kind of "reminder string"; *quipus* can describe history and non-numerical

facts. Quilter and Urton (2002) is an excellent reference for learning about *quipus* from an anthropological viewpoint.

- Details of 215 *quipus* can be found in Ascher and Ascher (1978, 1988). The Aschers' *Code of the Quipu Databook* is available on the Internet at http://courses.cit.cornell.edu/quipu/. A complete description on how to read the data book is also at the above site (and on the original or microfiche copy; see the Bibliography).
- Proposals of mathematical uses of the grid shown in Guamán Poma de Ayala's drawing, as well as the stone artifacts, have been given in Mackey et al. (1990) and in Higuera Acevedo (1994).
- Compare with the population of Otomi speakers in Chapter 9: The approximate number of Quechua speakers (according to Wikipedia) is about 6,000,000. Nevertheless, both Otomi and Quechua (as well as most indigenous languages) are strongly pressured to disappear or be reduced in number because of mainstream languages such as Spanish or English.

HINTS TO SELECTED EXERCISES

Important! Before you look at the hints and suggestions here, please take note of the following questions.

- *Always try to figure out the problem before looking at the hints.*
- *Many exercises have more than one solution or explanation. Some exercises are quite open-ended. Hence, your particular solution or explanation should be related to what is suggested here, but does not have to be exactly the same.*
- *If you have tried to solve the problem, have read these hints and suggestions, and still feel you cannot complete the exercise, do not give up! Talk it over with someone else, ask your instructor or someone familiar with the topic, or read parts or all of the chapter again. If you have the time, stop working on that exercise and go on to something else and come back to the exercise again later.*

Chapter 1

1.1 Pick up a newspaper, or read one on the Internet, and look for some of the ideas discussed in the chapter, such as an emphasis on number or assumptions of other cultures developing along the same path as has occurred in Western culture.

Introduction to Cultural Mathematics: With Case Studies in the Otomies and Incas, First Edition.
Thomas E. Gilsdorf.

1.4 **(a)** Focus on the word "adopted."

1.5 Focus on the word "precision."

1.7 There are a lot of numbers listed in the description, aren't there?

1.10 It is possible to have an even stronger emphasis on number.

1.11 Some partial information: *Measuring*: For the given data, we must calculate the size of the garden. Assuming a rectangular garden, we need four rows with 1 foot on either side of each row, for a total of 5 feet of width. Then we need 20 feet of length, so the area of the garden will be 100 square feet. The rectangular shape of the garden represents *location*. The final diagram of the garden arrangement is a case of *explaining*.

Explaining: The rows must be arranged in a certain order so that the plants will grow under optimal conditions. Thus, if we start with plant *A* first, then the second row cannot be plant *B*, so we choose either plant *C* or plant *D*. Suppose we choose plant *D* to go next to plant *A*. Then the third row cannot be plant *C*, so the only choice is plant *B*. There is also no other choice for the fourth row. You may wish to go through this reasoning with a different plant, say, plant *C*, in the first row.

Designing: We could develop a sort of algorithm for determining which rows go with which other rows. For example, because both plant *C* and plant *D* can be next to either plant *A* or plant *B*, we could assign some symbol to designate *either* plant *C* or plant *D*. Perhaps @ could be that symbol. Then we would have the structure of arranging plant *A* and plant *B* such that plant @ is between them.

Note: There is no rule that says a garden must be rectangular. Another choice is to have the garden be circular. You may wish to construct a diagram for the garden based on a circular shape.

1.13 This is a very important exercise. You may be surprised to find out that what people do in "ordinary" activities turns out to include mathematics! If you are not sure what to ask of the person you are interviewing, bring a copy of Bishop's six categories with you and start by asking what a "typical" day is like. See if you can make a connection with one of the six categories. Once you see one connection, it should be easier to find more.

1.14 This problem asks you to summarize the part of the chapter in which we described what "mathematics" means.

1.18 You may want to look up a reference on sociology or cultural anthropology, such as Robbins (2009). Ask yourself this question: How important is money in *your* life?

1.20 There are many possibilities for responding to this question. In the case of sewing a garment from a cloth, some partial information is this: *Designing*: You decide on an overall pattern, including colors, shapes, or details that will be part of the garment, and so on. *Measuring*: You measure many lengths (e.g., sleeves for a shirt). *Locating*: There are geometric concepts involved in putting together the overall pattern. *Explaining*: One could define some general terms such as what a "shirt" consists of, including ranges of lengths that correspond to particular sizes.

1.21 Start with the concept of "color."

Chapter 2

2.1 Look at a clock. Think about the kinds of numbers that are part of the answer to this question: How wide is a desk, table, chair, or other object where you are right now? There are other possibilities, too.

2.2 Go to a place that sells fruit and vegetables (e.g., a grocery store or supermarket) and think about how certain items are sold: Is it by the number of objects? If not, do you always know how many of those fruits or vegetables you are buying?

2.3 There are people from all kinds of cultures who have successfully obtained a Western education.

2.6 Here are three references that have many examples of different kinds of counting systems: Closs (1986), Zaslavsky (1999), and Barriga Puente (1998). The last one is in Spanish, but you can still figure out what some of the constructions are by looking at that reference. Another source would be to contact someone familiar with counting in other cultures, perhaps an elder of an indigenous culture, a cultural anthropologist, or a linguist. Here is an example of multiplication, so you can see how the description goes. From Barriga Puente (1998: 184–185): The number 2 is *Tiggene*, the number 10 is *Wàsha*, and the number 20 is *Tykeni d Washa*. We can infer that the structure of the number 20 comes from *Tykeni*, a variant of *Tiggene*, and *Washa*, a variant of *Wàsha*. Thus, 20 is 2×10.

2.7 **(a)** 25: Use *wikcémna nụ́p* to denote 20; then *sam*, which, if you look at Table 2.1, is used to connect the next value, then finish with the term for five, again from Table 2.1.

2.8 **(b)** 59: Look carefully at Table 2.3 and notice that when values are five or less than a multiple of 20, the structure is to express that value as a subtraction from that next multiple of 20. In the case of 59, it is one less than 60, so use the term *mọ́kòndín* (see, e.g., 39) to express the subtraction of 1, then the term for 60: *ádọta*. Remember to put the terms together as one word.

2.9 **(a)** *esaresa*: Notice how it has the term *esa*, which appears twice. It looks like two threes. Use this kind of reasoning for the other terms as well.

2.10 **(a)** *maseg scumu* contains the word for three and the word for four.

2.11 **(a)** The words for one through five appear to be independent. The word for six contains expressions for five and for one, so we can deduce that its structure is " 5 + 1" (but not "1 + 5"). This pattern continues for the next few values (but it's your job to do the describing, right?). The expression for ten contains elements of the expressions for five and two.

(b) In many cultures the expression for 20 has the form of something like "one complete person." What is the name of the culture for this problem?

(c) 19: The most likely structure would be that of 15 + 4, so you will have to explain these structures and how they would be connected to make 19. Note that order counts, too. For 35: Mimic the structure for 33. For 80: Look at the structure for 40.

2.13 This activity is probably best done with a group of several people (like a class of students). A discussion of the results and how they compare with how people from different places might perceive counting and numbers should give participants some insight into this chapter.

2.14 Some key points: As for common bases, think fingers, and possibly toes. Base 20 might have to do with humans having 20 fingers and toes (of course people of that particular group probably did not physically count their toes!). Once you realize the base 5 structure on your hands, you need not count toes directly. About an example, you may wish to look up number systems in Closs (1986), Barriga Puente (1998), or Zaslavsky (1999). About the cultural considerations: Consider that many cultures spend long periods of time in the same area, interacting with nearby cultures. Eventually, ideas and practices begin to mix between the cultures, including ways that people count. Another consideration is that a cultural group may have counted in a particular way early on, then modified its counting system over time. That culture may have retained some of its older counting system in the modified one. Such situations represent potential explanations for why a particular culture shows patterns of more than one base.

2.15 Some key points: Recall from Chapter 1 that distinct cultures can and may interpret mathematics very differently. A culture in which there is not an emphasis on number may have only a few number expressions in its language. On the other hand, people in that same culture may be very skilled at other mathematical tasks such as those described in Bishop's six categories. In later chapters we will see that kinship relations and activities such as creating certain kinds of art often include what we would consider complicated mathematics, independently of the kind of counting system used by that culture.

2.16 You should be able to find plenty of information about the Western counting system in books on the history of (Western) mathematics.

Chapter 3

3.1 The person can use one hand to express values from one to nine, so using two hands or gestures indicates something more.

3.3 If you have to count many things (say, more than about 25) by hand, how do you express the count?

3.4 See the discussion about making Egyptian symbols.

3.5 The commutative property was discussed in the text of the chapter. For the associative property (of addition; you may wish to write the explanation for the associative property of multiplication), you could describe, in words, that if you have three quantities, you could add the first two first, then add that subtotal to the third, or you could add the second two quantities first and add the first quantity to that subtotal, obtaining the same total value. (Note that this is not assuming the commutative property.) The point is that it does not matter at all *how* the quantities are expressed. With these comments, you can probably figure out how to explain the distributive property.

3.6 To imitate Western notation, you could define a rational number to be a ratio of integers (with the denominator not being zero). Notice that it does not matter how the numbers are expressed symbolically. For irrationals, think about how irrational numbers are defined in terms of rational numbers, and observe what, if anything, that has to do with symbolic representations of numbers.

3.8 Look for groups of ten symbols. For example, there are five 𓍢 symbols in each number, for a total of ten 𓍢. So, replace those ten with one symbol from the next higher power of ten. Think fish. Continue looking for ten with the other symbols from both numbers.

3.9 There are a total of eleven | symbols between both numbers, and that will get replaced by one ∩. Now, how many total ∩ symbols are there? Continue this line of thought for the rest of the symbols.

3.12 Recall that any time there are fewer symbols in the first number, you will have to replace a higher power with ten symbols in order to get enough extras in the first number (borrowing, if you want to call it that). There is only one 𓍢 in the first number and two 𓍢 in the second. Look at the next higher power of ten, namely, 𓆼, and replace one of them in the first number with ten 𓍢. That will give you a total of eleven 𓍢 in the first number, and two 𓍢 in the second number. The difference is nine 𓍢, but recall that now you have only two 𓆼 left in the first number. In you final answer, you will have those nine 𓍢. You will take two 𓆼 and subtract one, giving you one 𓆼 left. I think you can figure out the rest of the differences in symbols.

3.13 You will have to replace several symbols in the "borrowing" process. For example, with only one 𓎆 in the first number and two 𓎆 in the second, you will have to borrow from the next level. However, there are no 𓍢 in the first number, so borrow from the next following level, 𓆼. Replace one 𓆼 with ten 𓍢, then replace one of the ten 𓍢 with ten 𓎆, and carry on.

3.14 Follow the ideas of Example 3.4. To this end, note that 𓎆𓎆 times 𓎆𓎆 gives 𓍢𓍢𓍢𓍢, because each time we multiply a symbol by 𓎆 it gives us one symbol of the next level.

3.16 **(a)** Follow the ideas of Example 3.5a. For example, determine how many 𓎆 you would need to multiply to get one 𓆼, then how many 𓎆𓎆 are needed to get six 𓆼. Do the same comparison for the number of 𓎆𓎆 needed to get 𓎆𓎆𓎆𓎆. Put all the symbols together at the end.

(b) Follow the ideas of Example 3.5b, and part (a) above.

3.17 Observe that the question asks you to explain how the equation could be solved. If you want to solve the equation completely by using the symbols, follow the ideas of Example 3.6. You will have to explain how you have expressed fractions in the Egyptian symbols.

3.18 **(a)** For 8,000, use 八 and 千; for seven hundreds, use 七 and 百. A similar process will produce six tens. Finish with the symbol for two. As for whether the system is positional or not, consider what the value would be if you changed the order of the symbols.

(b) For the zero part, you just leave out the power of ten. For the other parts of the number, follow the ideas of the hints for part (a).

(c) (ii) There is no symbol for ten, but there is a symbol for 1,000, followed by a symbol for two. So, you have 1,000 and two. Note that if you reverse the order of the symbols, it gives you 2,000.

(iii) The trick is to organize the symbols according to powers of ten. About in the middle of the symbols is the one for 10,000: 萬.

This implies that the values that come before that symbol will multiply 10,000. The value of those symbols is given by looking at the group of symbols: 四千九百五十八, which represents four thousands, nine hundreds, five tens, and an eight. Multiply this value by 10,000, then figure out the value of the symbols that come after the one for 10,000. Put all the values together to get the final decimal value of the number.

(d) (ii) In the first number there is value of 47 for the last three symbols (a ten and a seven), and in the second number there is no symbol for ten, only a symbol for six. Because $47 + 6 = 53$, the last three symbols of the sum will look like this: 五十三 Continue this process of putting together symbols from each number for the next higher levels.

3.21 Each type of expression has its own advantages and disadvantages. For example, in a crowded, noisy room it might not be very efficient to communicate number information by writing them down, nor by shouting them out. You can reason through similar ideas for each of the three types of number expression.

3.23 Realize that a culture could be without a symbol for zero and still understand the concept involved. Many cultures that have not developed a writing system have various expressions for what it means to have "nothing" or even a specific word that means zero. Young children can often understand that if you have some objects and take all of them away, what remains is nothing, that is, zero, without using any written symbol for zero.

3.26 It may be useful for you to write out or explain some examples of your own, of how the Egyptians expressed numbers and number concepts. You may wish to consider that the symbols we looked at were used a long time ago by a culture that did not have the same priorities as might exist in your culture today. Use that difference of contexts to discuss the expression of numbers with symbols.

Chapter 4

4.2 **(c)** Keep in mind that our definition of cousin is a person that is a child of a sibling of one of your parents, but in the Warlpiri system, mothers and fathers come from different sections. There will be two possible sections here, but it is up to you to figure them out. Use the diagrams of the Warlpiri kinship system to help you out.

4.3 **(b)** This is a continuation of the ideas of Exercise 4.2. You will need to trace back through the previous generations of the girl, keeping in mind that mothers and fathers come from different sections. You may wish to make a diagram with lines connecting the ancestors of the girl. For example, you could start by letting G denote the girl. Then $m(G)$ is in section 2, while $f(G)$ is in 6. Then $mm(G)$ is in 4, and you can consider in which section is $mfm(G)$, and so on. To give you an idea, if you follow along only the matrilineal lines, you get this: $mmm(G)$ is in 1, and then $mmmm(G) = m^4(G)$ is back in 3. In this case, the girl's great-great-grandfather would be in section 7 because a woman in section 3 marries a man in section 7. Now consider other possibilities, such as $mfm(G)$, and following the generations back. Remember two things: (1) At each step, there is a mother and a father consideration that will lead you to different sections, and (2) the Warlpiri would not determine ancestry by making diagrams of section numbers as we are doing here.

4.6 **(b)** First, consider that a function is equal to itself except on the empty set, which is considered to be a finite set. For the transitive property, assume $f(x) = g(x)$ except on a finite set A, and $g(x) = h(x)$ except on a finite set B. You wish to show that the set of x-values on which $f(x) \neq h(x)$ must be finite, so consider what would be the the largest set of x-values for which $f(x) \neq h(x)$, with respect to the sets A and B.

4.8 For most of the parts of this question, keep in mind that $f^2 = I$, and $m^4 = I$. So, for example, $m^7 = m^4m^3 = m^3$. Thus, part (b) becomes $fm^3f = f(m^3f)$. Now look at Table 4.3 to determine which relation you get when you apply f to m^3f. Remember to apply f from the leftmost column to m^3f and not the other way around.

4.14 **(a)** The mother of the father is not the same as the father of the mother.

(b) Construct a table of operations similar to Table 4.2, consisting of three elements, say, a, b, and c. Decide which one of these will be the identity element, then fill in the table so that each element appears once and only once in each row and column. You will probably notice that there are not many ways to fill in the table. You will have to check to see that the associative property holds, but for operations with the identity element this is less work. When you are done, look at the final table. You will see that the operation must be commutative, so the group is abelian. As for the statement that every group of five or fewer elements must be abelian, the argument goes something like this: For four elements, you could construct a table of four elements as done above. It is more work than for three elements, but no matter how you construct it to obtain a group, the operation will be commutative. For five elements, there is a shorter route: Every group having a prime number of elements must be cyclic, and every cyclic group is abelian. This means that a group with five elements must be abelian.

Chapter 5

5.1 You can answer each part of this question in terms of concepts of symmetry. Keep in mind that "upside down" refers to reflecting across a horizontal line.

5.2 **(a)** Here is a clue for this question. Draw a square and notice that if you rotate it 90° by using a point in the center of the square, you will get back the same square. Now rotate it about a point that is not at the center, preferably far from it (such as near one of the vertices). Be careful rotating 90°. Do you get the same square back? Remember that the square must be in the same orientation (not tilted) in order for you to conclude "yes."

5.4 **(a)** If you draw a horizontal line through the center of the figure, do you get the same images on both the top and the bottom, with the bottom image in reverse orientation compared with the top? Repeat this thinking for vertical and rotational symmetry.

5.5 Keep in mind that the figure was made by painting and quill work and probably took a long time to make.

5.7 Be careful about conclusions regarding the symmetry of this figure. Notice that some parts of it overlap with others, so if you reflect the figure horizontally or vertically, it will look different. You can make statements about the general appearance of symmetry properties and clarify how the overlapping changes the symmetry properties.

5.10*–5.16* Hints for all of these exercises can be found in almost any textbook on the topic of abstract algebra (also called modern algebra). See Fraleigh (2003).

5.17 You will have to draw copies of the basic image in several directions, making the orientation go in the proper direction so as to achieve all of the symmetry properties.

5.19 If you just make a copy of the figure ♉ directly below (and oppositely oriented) from the original, then you will have 180° rotational symmetry and also horizontal symmetry. To not have the horizontal symmetry, think about a figure that has 180° rotational symmetry but not horizontal symmetry, such as the letter S. Notice where the bottom curve of the S is compared with the top curve.

5.21 The small circles do not exactly line up for the symmetry that appears to be satisfied.

Chapter 6

6.3 **(a)** Life is more complicated than this, isn't it?

(b) Calculate the total number of possibilities and ask yourself how easy it would be for a diviner to keep track of all of the information.

6.5 We have seen that the *ombiasy* is using the ideas of order of numbers (the order of the columns is important), parity (even versus odd), and operations that in Western terms would be referred to as Boolean algebra.

6.6 Think about the following: Suppose all four of the first columns are the same. What would that mean for all of the others? Now suppose that the first four columns are the same and are all o or all oo. What would these situations imply?

6.7 **(a)** $31 - 3 = 28 = 7 \times 4$, $18 - 2 = 16 = 4 \times 4$, 16 is divisible by 4, and 1 is less than four, so counts as 1. For the ordered pairs, remember that order is very important.

6.10 **(b)** $23 - 5 = 18 = 3 \times 6$, $33 - 3 = 30 = 5 \times 6$, and so on, using multiples of 6 in place of multiples of 4. For the ordered pairs, remember that order is important.

6.11 **(a)** For the first choice, there are 10 symbols total, and the diviner would choose two, so there are 10 possibilities for the first choice, and 10 for the second, for a total of 10×10. For the second part, there are three symbols and two are chosen, so you can repeat the thinking of the first part. The total will be the product of the two parts.

6.15 **(b)** Look back at the discussion regarding our proof that two columns must be equal.

6.16 **(a)** Write out the two possibilities for any element of $C_i + C_i$, and check them against $2 \times C_i$.

(c) Write three columns having distinct combinations that add up to oo in each entry of their sum.

6.19 The *random* part of the event now consists of three columns, so the total can be calculated by taking 2^4 to the third power. Notice that it does not matter how the other columns are created if they are all made using combinations of the original three randomly generated ones.

6.21* and **6.22*** For these questions, apart from the definition of congruence, you will need to recall that even numbers must be of the form $2k$ for some integer k, and odd numbers must be of the form $2k + 1$ for some integer k.

6.26 The *ombiasy* must have some assurance that nothing will appear in a tableau that would be impossible to interpret. Thus, properties that are crucial to the divination process must be valid for *any* possible tableau that could be created. This means the people who developed this kind of divination must have known that certain properties would be true for *any* final tableau. Such properties would include the property that at least two columns must be equal in the final tableau. We do not know what process was used to conclude such facts, but it must have been known.

6.27 A good source for finding out about other divination processes is Ascher (2002), especially the references at the end of chapter 1.

Chapter 7

7.1 **(b)** Card games are a good source. Chance describes what happens when players draw from the deck, and strategy describes how particular card combinations could be played. It is up to you to specify the game and details. If you do not play cards regularly, perhaps you are familiar with another game that includes chance and strategy, or you could look up how card games are played in the library or on the Internet.

(d) This is a difficult question!

7.2 Consider having the person cross with both the leopard and the cassava, for example.

7.5 Think about the strategy examples for the Hat Diviyan Keliya game, for example.

7.6 **(b)** Think about the man crossing the river with each object, and draw a diagram of what is left on one side of the river or the other, to see if it is possible to solve this one.

7.7 Think of vertices. Try to find the point on the game board for which the fewest number of leopards could be used to corner the tiger at that point. Any fewer leopards will be insufficient.

7.9 **(a)** There are 7! ways to arrange all seven coins, 2! ways to have two tails, and 5! ways to have the rest be heads. There are 2^7 possible outcomes. Now use the equations for N and P.

7.11 There are 15 total pieces, so think in terms of 6!, 3! (for the triangular pieces), and for the other nine, $P = 1/2^9$.

7.13 **(b)** The choice of which side would be considered "up" and which side would be "down" is arbitrary.

(c) This is a good example of the fact that, even when we might know what probable outcomes would be, we could include a special combination that gives a surprising outcome. In this case the surprising outcome is more points than another outcome that is equally probable. It adds excitement to the game.

7.16 If you wish, you can follow the form of description as I did for the Pay-gay-say game; however, you should feel free to write in your own style. How much mathematical detail you include depends on the details of the game and how in-depth your presentation or essay should be. Do not forget to include cultural aspects of the game you describe, for example, ceremonies that occur before, during, or after the game.

Chapter 8

8.2 Answer the question of: On which day does a new Nuu-chah-nulth year start?

8.3 **(a)** The winter solstice is on December 21, and the summer solstice is on June 21. To answer the question, think about what those two days are like.

(b) Think about part (a). Is there a way for a culture to change when those two days occur?

8.4 **(a)** A leap year is when the day of February 29 is added to the year after a certain number of years go by. To answer the second part of question, think about what would happen if no leap year were calculated for a very long time. Would it still be possible to make an adjustment? In what ways?

(b) Think about the Nuu-chah-nulth calendar.

8.7 This is related to Exercise 8.4 above. The Nuu-chah-nulth did not feel a need to adjust for leap year.

8.8 To answer the question, think about things such as: Is the Tzolkin cycle connected to a particular natural event or physical cycle?

8.9 Think about the importance of number and time in Western culture. Also, consider the assumption of other cultures following the same path of development as has occurred in Western culture. A review of Chapter 1 may be useful.

8.10 **(a)** Write using primes. You will need three threes, a two, and one 57. For parts (c) and (d) recall that when the cycles are independent, the supracycle is determined by the least common multiple, and if one cycle must finish before the other, then the supracycle is determined by the product of the numbers, regardless of common factors.

8.11 **(a)** There are approximately 52 weeks in a year: $52 \times 7 = 364$.

(b) After several years, 364 and 365.2422 will differ. Consider what would happen after 100 years.

(c) After some calculations, you should find that $255 = 7 \times 36 + 3$. If Tuesday represents the value one, then add three to it and determine which day of the week that comes to. For 171, the process is the same, but, of course, the value will be different.

8.12 **(a)** For each cycle, ask yourself questions such as those in Exercise 8.8.

(b) Think about how the end of one cycle and the beginning of the next cycle are determined.

8.13 You will need to calculate the least common multiple (*lcm*) of 360, 365, and 260. It will help to write the numbers in terms of their prime factors. Suppose now that the *lcm* turns out be the value $m \times 365$ (I am not going to tell you what m is because that would be giving away all the information!). Then that means the cycles coincide every m Vague Years. You can write your solution as m Vague Years. Repeat this process by writing the *lcm* as $n \times 360$ (n will not be the same value as m), and then do the same for 260 (call it k if you like, but the choice of letter is up to you). For solar years, you will have to write the *lcm* as a multiple of 365.2422, which of course will not be an integer. Nevertheless, you can still write the *lcm* as some number of these cycles, approximately.

8.15 **(a)** You will need to calculate $i_N = (12 + 631)(\text{mod } 13)$, which is $i_N = (643)(\text{mod } 13)$. Divide 631 by 13 and determine the remainder. Then you will have to calculate $j_N = (17 + 631)(\text{mod } 20)$. Notice that 660 is divisible by 20 and that $660 - 648 = 12$.

8.16 For a Tzolkin cycle to have both indices with the same value again, it would mean finding a value N such that $(j + N)(\text{mod } 13) = (j + N)(\text{mod } 20)$. This is a fancy way of stating that $j + N$ must be simultaneously divisible by 13 and 20. When would it be true that a number is divisible by both 13 and 20?

8.21 You will have to carry at the *Tun* and *Katun* levels.

8.22 You will have to carry at all levels except the first and last. Be careful with the value of eighteen at the *Uinal* level.

8.23 You will have to borrow at the *Kin* level. When you take out one point from the *Uinal*, it is equivalent to four bars, or twenty.

8.24 You will have to carry at several levels because of the zeros. Look at Example 8.19 for ideas, and remember that when you borrow from the *Tun* level for the *Uinal* level, it is a value of eighteen that is borrowed, not twenty.

8.31 Points that you would want to include would be that the Nuu-chah-nulth do not place the same importance on precision and number as is done in Western culture, but that their calendar served them well because as a culture, the Nuu-chah-nulth depended on natural cycles. Because natural cycles fluctuate, the Nuu-chah-nulth had to have flexibility built into it.

8.34 You will probably find that there are very few mathematical similarities between the ancient Egyptians and the Mayans.

Chapter 9

9.1 Think about the Otomi language compared with the Mayan and Aztec languages.

9.3 Think about Exercise 9.1 above, or look back to Chapter 2 on number systems.

9.4 **(b)** Assumes a similar path of development.

9.5 Ask yourself: What values of numbers are used for the day names in the Otomi calendar?

9.7 The Aztecs asked for many tributes of such items and seemed to keep very careful track of those items. What do you suppose that implies about how important they viewed those items?

9.8 **(a)** Ask yourself which cultural group had first settled in what is now central Mexico.

(b) Ask yourself which cultural group had a region-wide influence in the area and what that would imply.

9.10 First, you can write the value as: (6 × 100) + 20 + 10 + 5. Next, you will need to represent 600, which can be constructed by putting together *ʔrato* for six and *nthebe* for hundreds, followed by *ne* to connect hundreds with tens terms. Next, for thirty-five, recall that the Otomies would write this as "one twenty plus ten," so as we saw in Table 9.1 of the Otomi expressions, we have *ʔnate ma- ʔrœtʔa*, for thirty, and then you can add that to five, which is *kïtʔa*. Now put all these terms together. Note that the expression for thirty-five already appears somewhere in this chapter.

9.11 Follow the ideas of Example 9.2, changing the values in appropriate places.

9.14 How do you express fifty in Otomi? What value is *hñu*?

9.16 Notice that the expression *mahua,hi de ga mahua,hi* means "a thousand thousands" and is preceded by *kïtʔa*, which you can look up in the Otomi number systems section. Now note that *ʔrato* means six and is followed by *nthebe ne*, which implies multiplying the *ʔrato* value by 100 and then connecting it to the next lower term. Next, you will recognize the expression of "40 + 10," which, when connected with the previous value concerning *ʔrato*, is all multiplied by the term *nehua,hi*. This term is the version of 1,000 that is used for values over 100,000 (see, e.g., the expression for 800,000 in Table 9.1). Thus, for this part, we have 650,000. To finish, you will see the term *hñato*, followed by the last few terms of the expression. You can determine the values of these last values by using Table 9.1 and following the ideas of the three exercises preceding this exercise.

9.17 The number of cloths can be obtained by combining all symbols and noting that 20 "O" equals one flag.

9.18 **(a)** Ask yourself how many twenties there are in 400, and draw that many flags.

9.19 **(a)** Look at the examples of Aztec symbols in this chapter and rearrange them to see if that changes the values.

(b) Look back to Chapter 3 and the discussion of Egyptian symbols there.

9.20 Ask yourself questions such as whether you would get mirror images of the pattern by drawing a vertical line down the middle of the figure or whether you would get the same figure back if you rotated it 180°, as we did in Chapter 5 on art and decoration.

9.22 Note what happens if you try the glide reflect. Have you checked for other types of symmetry yet, such as vertical, horizontal, or 180° rotation?

9.24 You will need to solve the equations $i_M = (11 + 50)(\text{mod } 13) = 61(\text{mod } 13)$ and $j_M = (4 + 50)(\text{mod } 4) = 54(\text{mod } 4)$, as done in the examples.

9.26 Having several symmetry properties in artwork would imply a tendency to want order and organization.

9.28 If you read through the beginning of the chapter you will notice that the Otomies are an older group that are likely to have had some influence over the Aztecs when they first arrived; later when the Aztecs created their large empire, the influences were reversed. Try to incorporate these ideas into the topics of numeration, calendars, and so on.

9.30 The European view of women during the 1500s when the Spanish arrived in what is now Mexico was that work by women was considered as relatively unimportant. The important role of women as weavers in cultures such as the Otomies does not seem to be considered important by the Spanish. This gives you some ideas about how to explain the post-Conquest codex described in this question.

Chapter 10

10.1 If you put a sheet of paper near a window where sun enters, and place an object on it (that covers part but not all of the paper), then think about what happens to the color of the paper after a few days, compared with the color of the part of the paper that was covered. Now think about rocks sitting in the desert sun for hundreds of years and what the bottoms of those rocks would look like.

10.2 The question does not specify a human invasion. An invasion by much smaller, European-based life forms took place before human Europeans arrived. In fact, these smaller life forms proved to be catastrophically deadly to indigenous people of the Americas.

10.5 How educated, including in topics such as cultural anthropology and mathematics, were typical soldiers of that time period?

10.6 Guamán Poma de Ayala's book is a huge literary work that describes many aspects of Inca culture, including activities of people who were accountants and in similar positions. Thus, he must have known a lot about Inca culture. On the other hand, of the six or so drawings of *quipus* made by Guamán Poma de Ayala, only one shows some kind of knots (on which one cannot distinguish the knot structure), the others being left blank, and he did not include any description of how to interpret *quipus*.

10.7 Surely language is an obstacle; however, consider how a person of traditional Inca culture might respond to the following mathematics problem: If a person has six pennies in one pocket and seven pennies in another pocket, and wants to count how much money that is, how would that person count the money?

10.11 If you look at the figure again, you will notice that all of the values stated represent *pairs* of threads.

10.12 **(b)** The Incas counted in base 10. Ask yourself whether or not decimal counting is expressed in the figure in an obvious way.

10.13 One way could be to represent denominators using top cords. A denominator common to many values could be expressed as a dangle-end cord. Another way would be to construct cords that have two values, as we saw in Example 10.10. One value could represent the numerator value and the other could be the denominator value. One would need to be consistent; however, recall that in general the Incas were quite careful about spacing and organization in their *quipus*.

10.14 The knots on a *quipu* were tied in such a way that it would be difficult to untie and retie them. Do you remember how the knots were tied? Read the first few pages of the section on the *quipu* again if you need to refresh your memory.

10.15 Compare the words in the title of the chapter and the number expressions shown in Table 10.1. You should see part of a word that implies a number in Quechua. Also, recall the meaning of the word *Tawantinsuyu*.

10.16 **(a)** Here is a partial solution, in which I have left out the translations of some words.

$$[[[piqa]\ \{100\}\ (iskay)]\ (\{10\}\ (pusaq))]\ \{1000\}\ ([qanchis]\ \{10\}\ (tawa)).$$

(b) Now look up the translations of the values in part (a) such as *piqa* to finish.

10.18 **(a)** First write out the value as follows:

$$[1\times\{100\}+8\times\{10\}+2]\times\{1000\}+(7\times\{100\}+7\times\{10\}+4).$$

Now consult Table 10.1 for the expressions of the values as described above.

10.19 **(a)** Recall that the calculation involves the least common multiple, so determine $lcm(365, 12 \times 29) = lcm(365, 348)$. Notice that we multiply 29 by 12 for twelve lunar cycles. Now write each value using prime factors.

(c) Twelve lunar cycles of 29.5 days comes out to be 354. Proceed as in (a) and (b).

10.20 **(a)** Except for the sawtooth figures at the top of the design, there is vertical symmetry. If you include the sawtooth pattern, then none of the symmetry properties is satisfied.

(c) Refer to part (b) and observe that there does not seem to be a strong emphasis on even numbers, which we know the Inca culture has. This difference, although only in one example of Chimú art, indicates the possibility of cultural differences as expressed through mathematical properties.

10.22 **(a)** Check all three symmetry properties.

(b) Notice how the hourglass-type parts of the design represent pairs of smaller patterns.

10.23 **(a)** Before you start drawing, look at the values on all three cords, and observe that Cords 2 and 3 have values in the hundreds. What this tells you is that when you draw the knots for Cord 1, you will have to leave extra space for the hundreds place, leaving that space empty. Now, for Cord 1: Two simple knots (tens place), then eight loops, followed by four simple knots (tens place), and finally a figure eight knot. For Cord 2, put nine simple knots (hundreds place), then empty space in the tens area, followed by two loops; then for the second value, put five simple knots in the tens place followed by five loops. Proceed in the same way for Cord 3. At the end, the knots in the corresponding spaces should be carefully aligned. For instance, the five simple knots in the tens place of the second value on Cord 2 should be the same distance from the main cord as the four simple knots in the second value of Cord 1.

10.26 The problem is how to represent a person's name on a *quipu*. This is not as difficult as it seems. With a Western emphasis on number, many companies have constructed customer numbers to identify individual people. Compare this to businesses of a hundred years ago, when numerical data of customers was less detailed than it is now. Anyway, you will have to come up with a customer numbering scheme. It does not have to be complicated; the customer number can be assigned based on when the customer started buying from the company. The very first customer to buy something would be customer number one. This may not be practical for a large established company, but it is one possibility for you to think about. Similarly, you will have to come up with a way to assign numbers to street names that are not already numbered. Perhaps start with a number value for the first letter of the street name and then more values based on geographical information. For example, Central Avenue in a northern part of a particular city could be assigned a three (for "C" the third letter of the alphabet), followed by a "1" to represent that it is north of the city center (other values two through four would represent the other three cardinal directions, for instance), and some other number values that distinguish Central Avenue from other street names in the northern part of the city that also start with the letter "C." You can see that for this process of creating a *quipu*-based order form could get complicated depending on how many customers are involved and whether the company is local, regional, national, or international. The idea of this exercise is not to strive for complexity; keep the basic construction simple. Nevertheless, keep in mind that constructing such a numerical representation of data of individuals is not new. It is done all the time, on computers, in which, at the lowest level, all information is represented as binary (base 2) values!

10.33 Here are the cord values: Cord 1: 465; Cord 2: 701; Cord 3: 9; Cord 4: 531. It is up to you to draw the diagrams with the appropriate knots in the appropriate places, with the spacing organized well enough that there is no ambiguity in values (e.g., in values in the tens places versus hundreds places).

10.37 The main points of the essay or discussion should include descriptions of the following points: Much of our information comes from Spanish people who had early contact with the Inca. Early Spanish who occupied Inca territory probably had little education in topics such as cultural anthropology (which at the time did not even exist as a separate discipline) and mathematics. Also, the meanings of the terms "culture" and "mathematics" back then (1500s) were very different from how we are interpreting them here. Another aspect to include is that studies of people of Inca descent that have taken place in more recent times are studies of people who have been under the influence of non-Inca (i.e., Western) culture for hundreds of years, so observations about original Inca mathematics are difficult to conclude because of the mixing of cultures. Yet another aspect is that of archaeology: How archaeological artifacts are analyzed now is very different from how they were studied in the past. There are new techniques in archaeology that get developed every year and are more accurate in terms of identifying the age of objects, their possible uses, and so forth. Indeed, it is generally understood that many, perhaps hundreds, of *quipus* that were discovered starting with the Spanish occupation of Inca territory were discarded because their structure was not understood. Leland Locke's unraveling (no pun intended) of *quipu* structure did not become widely known until 1923, less than a hundred years ago. Finally, a discussion of obstacles to understanding Inca mathematics cannot finish without some mention of the problem of looting of archaeological sites. In Peru, for example, the looting of archaeological sites is a flourishing industry, with most of the artifacts ending up in places like the United States. Looted artifacts represent lost information that could otherwise have been useful to understanding the Incas and their mathematics.

10.42 Some points to consider: First, only one drawing of the dot diagram appeared in Guamán Poma de Ayala's work, and no explanation for it was given. Also, as noted earlier, of the six or so drawings of *quipus* made by him, only one shows some kind of knots (on which one cannot distinguish the knot structure), the others being left blank. Add to this the fact that Guamán Poma de Ayala did not describe any details of the structure of *quipus*. Thus, we must wonder how much he understood about *quipus* and their structure, and, likewise, how much he might have known about counting devices such as the possibility shown in the figure. Moreover, there has yet to be discovered a physical object that matches the diagram drawn by Guamán Poma de Ayala, and the stone structures that have been found (see Fig. 10.19) do not match the structure of Guamán Poma de Ayala's diagram, either.

BIBLIOGRAPHY

Acuña, René. 1990. *Arte breve de la lengua Otomi y vocabulario trilingüe, Español–Náhuatl–Otomi, de Alonso Urbano, 1605*. Reprinted by the National University of Mexico (UNAM), Mexico: UNAM.

Akar, Azade. 2004. *Traditional Turkish Designs, CD Rom and Book*. Mineola, NY: Dover Publications.

Albores, Beatriz. 2006. Los graniceros y el tiempo cósmico en la región que ocupó el Matlanzinco. In *Estudios de Cultura Otopame*, vol. 5. Mexico: National University of Mexico (UNAM) Instituto de Investigaciones, Antropológicas, 71–117.

Alcina Franch, José, and Josefina Palop Martínez. 1988. *Los Incas, el Reino del Sol*. Madrid: Ediciones Anaya.

Ancient Mexican Designs, CD Rom and Book. 2002. Mineola, NY: Dover Publications, Dover Electronic Clip Art.

Appelbaum, Stanley, ed. 2006. *Treasury of Chinese Designs, CD Rom and Book*. Mineola, NY: Dover Publications,.

Appleton, Le Roy. 2003. *Appleton's American Indian Designs, CD Rom and Book*. Mineola, NY: Dover Publications.

Artes de México. 2000. *Textiles de Oaxaca*, 2nd edition, no. 35.

Ascher, Marcia. 1983. "The logical-numerical system of Inca quipus." *Annals of the History of Computing*, 5(3):268–278.

Ascher, Marcia. 1986. "Mathematical ideas of the Incas." In *Native American Mathematics*, ed. Michael P. Closs. Austin: University of Texas Press, 261–289.

Introduction to Cultural Mathematics: With Case Studies in the Otomies and Incas, First Edition.
Thomas E. Gilsdorf.

Ascher, Marcia. 1991. *Ethnomathematics: A Multicultural View of Mathematical Ideas*. Pacific Grove, CA: Brooks/Cole.

Ascher, Marcia. 2002. *Mathematics Elsewhere*. Princeton, NJ: Princeton University Press.

Ascher, Marcia, and Robert Ascher. 1978. *Code of the Quipu: Databook*. Ann Arbor: University of Michigan Press.

Ascher, Marcia, and Robert Ascher. 1981. *Code of the Quipu: A Study in Media, Mathematics, and Culture*. Ann Arbor: University of Michigan Press (reprinted as *Mathematics of the Incas: Code of the Quipu*, Dover Publications, 1997).

Ascher, Marcia, and Robert Ascher. 1988. *Code of the Quipu: Databook II*. Ithaca, NY: Ascher and Ascher (on microfiche at Cornell University Archives, Ithaca, NY).

Ascher, Marcia, and Ubiratan D'Ambrosio. 1994. "Ethnomathematics: a dialogue." *For the Learning of Mathematics* 14(2):36–43.

Atwood, Roger. 2003. "Guardians of the dead." *Archaeology* January/February: 42–49.

Aveni, Anthony, et al., eds. 1990. *The Lines of Nazca*. Philadelphia: American Philosophical Society, Memoirs, vol. 183.

Bajot, Édouard. 2006. *French Decorative Designs of the 18th Century: CD Rom and Book*. Mineola, NY: Dover Publications.

Barriga Puente, Francisco. 1998. *Los sistemas de numeración indoamericanos: un enfoque areotipológico*. Mexico City: UNAM.

Bartholomew, Doris. 2000. Intercambio lingüístico entre Otomi y Náhuatl. In *Estudios de Cultura Otopame*, vol. 2, ed. Y. Lastra and N. Quezada. Mexico City: UNAM, 189–201.

Beeler, Madison S. 1986. "Chumash numerals." In *Native American Mathematics*, ed. Michael P. Closs. Austin: University of Texas Press, 109–128.

Belanger Grafton, Carol, ed. 2008. *Pre-Columbian Mexican Designs: CD Rom and Book*. Mineola, NY: Dover Publications, Dover Electronic Clip Art.

Belcastro, Sarah-Marie, and Carolyn Yackel, eds. 2011. *Crafting by Concepts: Fiber Arts and Mathematics*. Boca Raton, FL: CRC Press.

Bernal Perez, Felipino. 2001. *Diccionario Hñähñu–Español, Español–Hñähñu del Valle del Mezquital Hidalgo, Tercera Ed*. Ixmiquilpan, Mexico: Hmunts'a Hemi.

Bishop, Alan J. 1979. "Visualising and mathematics in a pre-technological culture." *Educational Studies in Mathematics* 10:135–146.

Bishop, Alan J. 1988. *Mathematical Enculturation: A Cultural Perspective on Mathematics Education*. Dordrecht: Kluwer Academic Publishers.

Broda de Casas, Johanna. 1969. *The Mexican Calendar as Compared to Other Mesoamerican Systems*. Acta Ethnologica et Linguistica, Series Americana, vol. 4. Wien: Institut für Völkerkunde der Universität Wien.

Broda de Casas, Johanna. 1991. *Arqueoastronomía y etnoastronomía en Mesoamérica*. Mexico City: UNAM.

Burton, David M. 2007. *The History of Mathematics: An Introduction*. Boston: McGraw-Hill.

Cahlander, Adele. 1985. *Double-Woven Treasures from Old Peru* (with Suzanne Baizerman). St. Paul, MN: Dos Tejedoras Publications.

Calinger, Ronald. 1999. *A Contextual History of Mathematics, to Euler*. Upper Saddle River, NJ: Prentice Hall.

"Caroline Islands." 2008. *The Columbia Encyclopedia*, 6th edition. Available at: http://www.encyclopedia.com (accessed September 2, 2009).

Carrasco Pizana, Pedro. 1950. *Los Otomíes*. Mexico City: UNAM, Instituto de Historia.

Caso, Alfonso. 1967. *Los calendarios prehispánicos*. Mexico City: UNAM, Instituto de Investigaciones Históricas.

Cason, Marjorie, and Adele Cahlander. 1976. *The Art of Bolivian Highland Weaving*. New York: Watson-Guptill Publications.

Castillo Escalona, Aurora. 2000. *Persistencia histórico-cultural San Miguel Tolimán*. Querétero: Universidad Autónoma de Querétero.

Chemín Bässler, Heidi. 1993. *Las capillas oratorio otomíes de San Miguel Tolimán: Ya t'u̲lo Nijõ dega södi ñuhu ya menga Nxe̲mge̲*. Querétero: Fondo Editorial de Querétero.

Cirillo, Dexter. 1998. *Across Frontiers: Hispanic Crafts of New Mexico*. San Francisco, CA: Chronicle Books.

Closs, Michael P., ed. 1986. *Native American Mathematics*. Austin: University of Texas Press.

Cullen, Stewart. 1975. *Games of the North American Indians*. New York: Dover Publications.

D'Addetta, Joseph. 2007a. *American Folk Art Designs, CD Rom and Book*. Mineola, NY: Dover Publications.

D'Addetta, Joseph. 2007b. *Japanese Motifs and Designs, CD Rom and Book*. Mineola, NY: Dover Publications.

D'Altroy, Terence. 2003. *The Incas*. Oxford: Blackwell Publishing.

D'Ambrosio, Ubiratan. 1985. "Ethnomathematics and its place in the history and pedagogy of mathematics." *For the Learning of Mathematics* 5(1):44–48.

D'Ambrosio, Ubiratan. 2000. *Educaçao Matemática, da Teoria à Prática*. Papirus Editora Brazil.

D'Ambrosio, Ubiratan. 2006. *Ethnomathematics*. Boston: Sense Publishers.

Darling, David. 2009. "Mathematics as a universal language." Available at: http://www.daviddarling.info/encyclopedia/M/math.html (accessed July 7, 2009).

Dauben, Joseph. 2007. "Chinese mathematics." In *The Mathematics of Egypt, Mesopotamia, China, India, and Islam: A Sourcebook*, ed. Victor Katz. Princeton, NJ: Princeton University Press, 187–384.

Denny, J. Peter. 1986. "Cultural ecology of mathematics: Ojibway and Inuit hunters." In *Native American Mathematics*, ed. Michael Closs. Austin: University of Texas Press, 129–180.

Dow, James. 1996. "Ritual presentation, intermediate-level social organization, and Sierra Otomí oratory groups." *Ethnology* 35(3):195–202.

Dowlatshahi, Ali. 2007. *Persian Designs and Motifs, CD Rom and Book*. Mineola, NY: Dover Publications.

Eglash, Ron. 1999. *African Fractals: Modern Computing and Indigenous Design*. New Brunswick, NJ: Rutgers University Press.

Enciso, Jorge. 2004. *Design Motifs of Ancient Mexico*. Mineola, NY: Dover Publications.

Eves, Howard. 1990. *An Introduction to the History of Mathematics with Cultural Connections* (by Jamie H. Eves), 6th Ed. Pacific Grove, CA: Brooks/Cole.

Fiadone, Alejandro Eduardo. 1997. *Argentine Indian Art*. Mineola, NY: Dover Publications.

Folan, William J. 1986. "The calendrical and numerical systems of the Nootka." In *Native American Mathematics*, ed. Michael P. Closs. Austin: University of Texas Press, 93–107.

Fraleigh, John B. 2003. *A First Course in Abstract Algebra*, 7th edition. Boston: Addison Wesley, 2003.

Freund, John. 1993. *Introduction to Probability*. Mineola, NY: Dover Publications.

Galinier, Jacques. 1990. *La mitad del mundo: cuerpo y cosmos en lo rituales otomíes*. Mexico City: UNAM, 1990.

Gallian, Joseph A. 1990. *Contemporary Abstract Algebra*, 2nd edition. Toronto: D.C. Heath.

Gerdes, Paulus. 1998. *Women Art and Geometry in Southern Africa*. Trenton, NJ: Africa World Press.

Gerdes, Paulus. 2007. *Drawings from Angola: Living Mathematics*. Raleigh, NC: Lulu Press.

Gerdes, Paulus. 2008. *African Basketry: A Gallery of Twill-Plaited Designs and Patterns*. Raleigh, NC: Lulu Press.

Gillings, Richard J. 1981. *Mathematics in the Time of the Pharoahs*. Dover Publications.

Gilsdorf, Thomas E. 2000. "Inca Mathematics." In *Mathematics Across Cultures: The History of Non-Western Mathematics*, ed. Helaine Selin. Dordrecht: Kluwer Academic Publishers, 189–203.

Gilsdorf, Thomas E. 2008a. "Ethnomatemáticas de los Otomíes." *Estudios de Cultura Otopame*, vol. 6, Mexico City: UNAM, 165–181.

Gilsdorf, Thomas E. 2008b. "Ethnomathematics of the Inkas." In *Encyclopaedia of the History of Science, Technology, and Medicine in Non-Western Cultures*, 2nd edition (electronic), ed. Helaine Selin. New York: Springer-Verlag.

Gilsdorf, Thomas E. 2009. "Mathematics of the Hñähñu: The Otomies." *Journal of Mathematics and Culture* 4(3): 84–105.

Green, Miranda. 1996. *The Gods of the Celts*. Totowa, NJ: Barnes and Noble.

Guaman Poma de Ayala, Felipe. 1980. *El primer nueva corónica y buen gobierno* (1614). Mexico City: Siglo XXI Editores.

Guerrero Guerrero, Raúl. 1983. *Los Otomíes del Valles del Mezquital*. Pachuca: INAH.

d'Harcourt, Raoul. 2002. *Textiles of Ancient Peru and their Techniques*. Mineola, NY: Dover Publications.

Hays, Wilma, and R. Vernon Hays. 1973. *Foods the Indians Gave Us*. Philadelphia: David McKay.

Hecht, Ann. 1989. *The Art of the Loom: Weaving, Spinning and Dyeing across the World*. New York: Rizzoli.

Hermann Lejarazu, Manuel. 2001. "Códices tributaries de Mezquiahuala." In *Códices del Estado de Hidalgo, State of Hidalgo Codices*, ed. Laura Elena Sotelo Santos, Victor Manuel Ballesteros García, and Evaristo Luvián Torres. Pachuca: UAEH, 88–99.

Higgens, Charlotte. 2010. "Winner announced for world's oddest book title award." *The Guardian*, March 26.

Higuera Acevedo, Clara L. 1994. "La yupana Incaica: Elemento histórico como instrumento pedagógico." In *Proceedings of the Meeting of the International Study Group on Relations between History and Pedagogy of Mathematics*, ed. Sergio Nobre. Brazil: UNESP, 77–89.

Imhausen, Annette. 2007. "Egyptian mathematics." In *The Mathematics of Egypt, Mesopotamia, China, India, and Islam: A Sourcebook*, ed. Victor Katz. Princeton, NJ: Princeton University Press, 7–56.

Joseph, George Gheverghese. 2011. *The Crest of the Peacock: Non-European Roots of Mathematics*, 3rd edition. Princeton, NJ: Princeton University Press.

Katz, Victor, ed. 2007. *The Mathematics of Egypt, Mesopotamia, China, India, and Islam: A Sourcebook*. Princeton, NJ: Princeton University Press.

Lara, Jesus. 1960. *La Literatura de los Quechuas*. Bolivia: Editorial Canelas.

Lastra, Yolanda. 2006. *Los Otomíes: Su Lengua y su Historia*. Mexico: Universidad Nacional Autónoma de México-Instituto de Investigaciones Antropológicas.

Lastra, Yolanda, and Doris Bartholomew, eds. 2001. *Códice de Huichapan: paleografía y traducción*. Translation of notes by Lawrence Ecker. Mexico City: Instituto de Investigaciones Antropológicas, UNAM.

Levin, Norman Balfour. 1964. *The Assiniboine Language*. Bloomington: Mouton & Co. La Haya.

Locke, Leland L. 1912. "The ancient Quipu or Peruvian knot record." *American Anthropologist*, New Series, 14(2): 325–332.

Lowery, Philip, and Richard Savage. 1976. "Celtic design with compasses as seen on the Holcombe Mirror." In *Celtic Art in Ancient Europe: Five Protohistoric Centuries*, ed. Paul-Marie Dural and Christopher Hawkes. London: Seminar Press, 219–231.

Luces Contemporaneas del Otomí. 1979. Mexico City: Instituto Lingüístico de Verano.

Lusebrink, Amy, and Courtney Davis. 2006. *Celtic and Norse Designs, CD Rom and Book*. Mineola, NY: Dover Publications.

Mackey, Carol, Hugo Pereyra, Carlos Radicati, Humberto Rodriguez, and Oscar Valverde, eds. 1990. *Quipu y Yupana, Colección de Escritos*. Lima: CONCYTEC.

Mannheim, Bruce. 1985. "Southern Peruvian Quechua." In *South American Indian Languages, Retrospect and Prospect*, ed. Harriet E. Manelis and Louisa S. Stark. Austin: University of Texas Press, 481–515.

Mannheim, Bruce. 1991. *The Language of the Inka since the European Invasion*. Austin: University of Texas Press.

Menzel, Dorothy. 1976. *Pottery Style and Society in Ancient Peru*. Berkeley: University of California Press.

Merrifield, William R. 1968. "Number names in four languages of Mexico." In *Grammars for Number Names*, ed. H. Brandt Corstius. Dordrecht: D. Reidel.

Milbrath, Susan. 1999. *Star Gods of the Maya: Astronomy in Art, Folklore, and Calendars*. Austin: University of Texas Press.

Moseley, Michael E. 1992. *The Incas and Their Ancestors: The Archeology of Peru*. London: Thames and Hudson.

Murray, H.J.R. 1952. *The History of Board Games Other Than Chess*. Oxford: Oxford University Press.

Naylor, Maria, ed. 1975. *Authentic Indian Designs*. Mineola, NY: Dover Publications.

Nelson, Richard. 1993. "Understanding Eskimo science." In *Annual Editions: Anthropology, 2008– 2009*. Boston: McGraw-Hill, 68–70.

Niethammer, Carolyn. 1974. *Indian Food and Lore: 150 Authentic Recipes*. New York: Macmillan.

Norton-Smith, Thomas M. 2004. "Indigenous numerical thought in two American Indian tribes." In *American Indian Thought*, ed. Anne Waters. Malden, MA: Blackwell, 58–71.

Orban-Szontagh, Madeleine. 1993. *261 North American Indian Designs*. Mineola, NY: Dover Publications.

Pacheco, Oscar. 1997. *Ethnogeometría*. Santa Cruz (Bolivia): Editorial CEPDI Bolivia.

Partlett, David. 1999. *The Oxford History of Board Games*. Oxford: Oxford University Press.

Pasquaretta, Paul. 2003. *Gambling and Survival in Native North America*. Tucson: University of Arizona Press.

Paternosto, César. 1996. *The Stone and the Thread: Andean Roots of Abstract Art*. Translated by Esther Allen. Austin: University of Texas Press.

Pinxten, R., van Dooren, I., and F. Harvey. 1983. *The Anthropology of Space*. Philadelphia: University of Pennsylvania Press.

Pollard Rowe, Ann. 1977. *Warp Patterned Weaves of the Andes*. Washington: Textile Museum.

Powell, Arthur B., and Marilyn Frankenstein, eds. 1997. *Ethnomathematics: Challenging Eurocentrism in Mathematics Education*. Albany: State University of New York Press.

Quilter, Jeffrey and Urton, Gary, eds. 2002. *Narrative Threads: Accounting and Recounting in Andean Khipu*. Austin: University of Texas Press.

Radicati Di Primeglio, Carlos. 1973. *El sistema contable de los Incas, Yupana y Quipu*. Lima: Librería Studium.

Reagan, Albert B., and F. W. Waugh. 1919. "Some games of the Bois Fort Ojibwa." *American Anthropologist*, New Series, 21(3) (July–September):264 –278.

Reiche, Maria. 1993. *Contribuciones a la Geometría y Astronomía en el Antiguo Perú*. Lima: Epígrafe Editores.

Robbins, Richard H. 2009. *Cultural Anthropology: A Problem Based Approach*, 5th edition. Belmont, CA: Wadsworth Cengage Learning.

Rosenthal, Elisabeth. 2009. "Students give up wheels for their own two feet." *New York Times*, March 26. Available at: http://www.nytimes.com/2009/03/27/world/europe/27bus.html?_r=1&scp=1&sq=Students%20Give%20Up%20Wheels%20for%20Their%20Own%20Two%20Feet&st=cse.

Ross, Kurt. 1978. *Codex Mendoza, Aztec Manuscript*. CH-Fribourg: Miller Graphics/Liber S.A.

Sandstrom, Alan R. 1981. *Traditional Curing and Crop Fertility Rituals Among Otomi Indians of the Sierra de Puebla, Mexico: The Lopez Manuscripts*. Bloomington: Indiana University Press.

Sayer, Chloë. 1990a. *Arts and Crafts of Mexico*. San Francisco: Chronicle Books.

Sayer, Chloë. 1990b. *Mexican Patterns, A Design Source Book*. New York: Portland House.

Schneider, Mary Jane. 1994. *North Dakota Indians: An Introduction*. Dubuque, IA: Kendall/Hunt.

Selin, Helaine, ed. 2000a. *Astronomy Across Cultures: A History of Non-Western Astronomy*. Dordrecht: Kluwer Academic Publishers.

Selin, Helaine, ed. 2000b. *Mathematics Across Cultures: The History of Non-Western Mathematics*. Dordrecht: Kluwer Academic Publishers.

Selin, Helaine, ed. 2008. *Encyclopaedia of the History of Science, Technology, and Medicine in Non-Western Cultures*. Berlin: Springer-Verlag.

Sentilles, Dennis. 1975. *A Bridge to Advanced Mathematics*. Baltimore, MD: Williams & Wilkins.

Shaffer, Frederick W. 1979. *Indian Designs from Ancient Ecuador*. Mineola, NY: Dover Publications.

Smith, D. E. 1923. *History of Mathematics*. Reprinted by Dover Publications, 1958.

Smith, Michael. 2003. *The Aztecs*, 2nd edition. Malden, MA: Blackwell Publishers.

Soustelle, Jacques. 1937. *La famille Otomi-pame du Mexique Central*, Travaux en Mémoires de l'Institut d'Ethnologie, Paris. Translation into Spanish by Centro de Estudios Mexicanos y Centroamericanos, Fondo de Cultura Económica, 1993.

Stark, Louisa R. 1985a. "Ecuadorian highland Quechua: history and current status." In *South American Indian Languages, Retrospect and Prospect*, ed. Harriet E. Manelis and Louisa S. Stark. Austin: University of Texas Press, 443–479.

Stark, Louisa R. 1985b. "The Quechua language in Bolivia." In *South American Indian Languages, Retrospect and Prospect*, ed. Harriet E. Manelis and Louisa S. Stark. Austin: University of Texas Press, 516–545.

Taimina, Daina. 2009. *Crocheting Adventures with Hyperbolic Planes*. Wellesley, MA: AK Peters Publishing.

Thomas, Cyrus. 1897. "Numeral systems of Mexico and Central America." In *Nineteenth Annual Report*, Bureau of American Ethnology. Washington, DC.

Turner, Wilson G. 1980. *Maya Designs*. Mineola, NY: Dover Publications.

Turner, Wilson G. 2005. *Aztec Designs*. Mineola, NY: Dover Publications.

Urton, Gary. 1981. *At the Crossroads of the Earth and Sky: An Andean Cosmology*. Austin: University of Texas Press.

Urton, Gary. 1997. *The Social Life of Numbers: A Quechua Ontology of Numbers and Philosophy of Arithmetic*. Austin: University of Texas Press.

Urton, Gary. 1999. *Inca Myths*. Austin: University of Texas Press.

Urton, Gary. 2003. *Signs of the Inka Khipu: Binary Coding in the Andean Knotted String Records*. Austin: University of Texas Press.

Valiñas Coalla, Leopoldo. 2000. "Lo que la lingüística yutoazteca podría aportar en la reconstrucción histórica del Norte de México." In *Nómada y Sedentario en el Norte de México, Homenaje a Beatriz Branniff*, ed. Marie-Areti Hers, José Luís Mirafuentes, María de los Dolores Soto, and Miguel Vallebueno. Mexico City: UNAM, 175–205.

Valle, Perla. 2001. "Códice osuna." In *Códices del Estado de Hidalgo, State of Hidalgo Codices*, ed. Laura Elena Sotelo Santos, Victor Manuel Ballesteros García, and Evaristo Luvián Torres. Pachuca: UAEH, 56–63.

Ventura, Carol. 2003. *Maya Hair Sashes Backstrap Woven in Jacaltenango, Guatemala*, 2nd edition. *Cintas Mayas Tejidas con el Telar de Cintura en Jacaltenango, Guatemala*. Cookeville: CV.

Verran, Helen. 2000. "Accounting mathematics in West Africa: some stories of Yoruba number." In *Mathematics Across Cultures: A History of Non-Western Mathematics*, ed. Helaine Selin. Dordrecht: Kluwer Academic Publishers, 345–371.

Wasserman, Tamara E., and Jonathan S. Hill. 1981. *Bolivian Indian Textiles, Traditional Designs and Costumes*. New York: Dover Publications.

Waters, Anne, ed. 2004. *American Indian Thought*. Oxford: Blackwell Publishing.

Weber, John David. 1989. *A Grammar of Huallaga (Huánuco) Quechua*. Berkeley and Los Angeles: University of California Press.

Weitlaner-Johnson, Irmgard. 1993. *Mexican Indian Folk Designs, 252 Motifs from Textiles*. Mineola, NY: Dover. Publications.

Weller, Alan. 2008. *Pre-Columbian Design*. Mineola, NY: Dover Publications.

Williams, Geoffrey, ed. 2004. *African Tribal Designs, CD Rom and Book*. Mineola, NY: Dover Publications.

Wilson, Eva. 1984. *North American Indian Designs for Artists and Craftspeople*. Mineola, NY: Dover Publications.

Womack, Mari. 2010. *The Anthropology of Health and Healing*. Lanham, MD: Altamira Press.

Wood, Leigh N. 2000. "Communicating mathematics across culture and time." In *Mathematics Across Cultures: The History of Non-Western Mathematics*, ed. Helaine Selin. Dordrecht: Kluwer Academic Publishers, 1–12.

Wright, David. 1997. Manuscritos Otomies del Virreinato. In *Códices y Documentos sobre México, Segundo Simposio*, Volumen II, ed. Salvador Rueda Mithers, Constanza Vega Sosa, and Rodrigo Martínez Barca. Mexico City: Instituto Nacional de Antropología e Historia (INAH), 437–462.

Wright, David. 2005. "Hñahñu, Ñuhu, Ñhato, Ñuhmu." *Arqueología Mexicana* 13(73) (May–June):19.

Yam, Emilie. 2009. "Young Americans going abroad to teach." CNN.com, March 26. Available at: http://www.cnn.com/2009/TRAVEL/03/20/teaching.abroad/.

Zaslavsky, Claudia. 1995. *The Multicultural Classroom: Bringing in the World*. Portsmouth, NJ: Heinemann.

Zaslavsky, Claudia. 1998. *Math Games and Activities from Around the World*. Chicago, IL: Chicago Review Press.

Zaslavsky, Claudia. 1999. *Africa Counts: Number and Pattern in African Culture*, 3rd edition. Chicago, IL: Lawrence Hill Books.

Zaslavsky, Claudia. 2003. *More Math Games and Activities from Around the World*. Chicago, IL: Chicago Review Press.

INDEX

Introduction to Cultural Mathematics: With Case Studies in the Otomies and Incas, First Edition. Thomas E. Gilsdorf.